Sustainable Supply Chains

International Series in Operations Research & Management Science

Volume 174

For further volumes:
http://www.springer.com/series/6161

Tonya Boone • Vaidyanathan Jayaraman
Ram Ganeshan
Editors

Sustainable Supply Chains

Models, Methods, and Public Policy Implications

Springer

Editors
Tonya Boone Ph.D.
Mason School of Business
College of William and Mary
Williamsburg, Virginia
USA

Ram Ganeshan Ph.D.
Mason School of Business
College of William and Mary
Williamsburg, Virginia
USA

Vaidyanathan Jayaraman Ph.D.
Department of Management
School of Business Administration
University of Miami
Coral Gables, Florida
USA

ISSN 0884-8289
ISBN 978-1-4419-6104-4   ISBN 978-1-4419-6105-1 (e-Book)
DOI 10.1007/978-1-4419-6105-1
Springer New York Dordrecht Heidelberg London

Library of Congress Control Number: 2012939732

Printed on acid-free paper

Springer is part of Springer Science+Business Media (www.springer.com)

Acknowledgements

The editors would like to extend their heartfelt gratitude to the Mason School of Business, College of William & Mary; and the School of Business, University of Miami, for supporting their research endeavors in the writing of this book.

Tonya Boone and Ram Ganeshan would especially like to thank Dean Ajit Rangnekar, Senior Associate Dean Sanjay Kallapur, colleagues and friends at the Indian School of Business in Hyderabad for their support during their sabbatical when much of the finishing touches to this book were completed.

All the three editors are extremely grateful to their families for inspiring them with wonderful ideas on sustainability.

We are grateful to our editor Neil Levine and assistant editor Matthew Amboy for sheparding this book through the publication process.

Finally, we would also like to thank you, our readers, for reading and using this book. We hope it contributes to all the efforts that you've undertaken to improve the sustainable performance of companies and their supply chains throughout the world.

Contents

List of Contributors

Ajay Anandnarayan Business Consultant, Cognizant Business Consulting, Chennai, India

Paul M. Ayoub Shell Alternative Energies, One Shell Plaza, Houston, TX 77002, USA

Chunguang Bai School of Management Science and Engineering, Dongbei University of Finance and Economics, 217 Jianshan Street, 116025 Dalian, People's Republic of China

Grahame Y. Buss Shell Global Solutions, London, UK

Tonya Boone Mason School of Business, College of William and Mary, 23185 Williamsburg, VA, USA

Gökçe Esenduran Fisher College of Business, The Ohio State University, 2100 Neil Avenue, Columbus 43210, OH, USA

Mark Ferguson Moore School of Business, University of South Carolina, 1705 College Street, Columbia, SC 29208, USA

Ram Ganeshan Mason School of Business, College of William and Mary, 23185 Williamsburg, VA, USA

Robert L. Hicks 129 Morton Hall, Department of Economics, The College of William and Mary, Williamsburg 23185, VA, USA

Rex Holmlin Mason School of Business, The College of William and Mary, 101 Ukrop Way,Williamsburg, VA 23185, USA

Vaidyanathan Jayaraman Department of Management, School of Business Administration, University of Miami, 33124 Coral Gables, FL, USA

Sachin Kapila Shell Global Solutions, London, UK

Eda Kemahlioğlu-Ziya Kenan-Flagler Business School, University of North Carolina at Chapel Hill, CB Number 3490, Chapel Hill 27599, NC, USA

Mandar Dattatray Malshe Business Analyst (Consumer Packaged Goods), Wipro Technologies, Chennai, India

Amir H. Masoumi Department of Finance and Operations Management, Isenberg School of Management, University of Massachusetts, 01003 Amherst, MA, USA

Anna Nagurney Department of Finance and Operations Management, Isenberg School of Management, University of Massachusetts, 01003 Amherst, MA, USA

Alex P. R. Nevill Shell Alternative Energies, Shell International Petroleum Company Ltd, London, UK

Joseph B. Powell Shell Projects and Technology, Westhollow Technology Center, Houston, TX 77082, USA

Joseph Sarkis Graduate School of Management, Clark University, 950 Main Street, 01610-1477, Worcester, MA, USA

Gilvan C. Souza Kelley School of Business, Indiana University, 47405 Bloomington, IN, USA

Ravi Subramanian College of Management, Georgia Institute of Technology, Atlanta, GA 30308, USA

Jayashankar M. Swaminathan Kenan-Flagler Business School, University of North Carolina at Chapel Hill, CB Number 3490, Chapel Hill 27599, NC, USA

Chapter 1
Editorial: Sustainable Supply Chains

Tonya Boone, Vaidyanathan Jayaraman and Ram Ganeshan

Abstract This editorial serves as an introduction to this book—it provides a brief overview of sustainability and traces the evolution of sustainability and sustainable supply chains over the last fifty years.

1.1 Overview

The Bruntland commission of the United Nations defined sustainability and sustainable development as:

> ... sustainable development is development that meets the needs of the present without compromising the ability of future generations to meet their own needs.

Malthus predicted that the growth in the worlds' population would have catastrophic consequences—while this prediction has not yet come true, concern definitely exists that rapid population growth will have substantial negative social and environmental consequences[1]. Today there are about 7 billion people in the world out of which around 1.5 billion live in conditions of abject poverty. As many as 800 million people are severely malnourished. Environmental degradation due to human activities continues to progress at an increasing rate with annual emissions of carbon-dioxide having increased by nearly 80% between 1970 and 2004 (IPCC 2007).

Chris Field, Director of the Carnegie Institution's Department of Global Ecology and co-chair of the Intergovernmental Panel on Climate Change's (IPCC) Working Group 2 recently said:

[1] Malthus TR. *An essay on the principle of population* (1798). Ed. P. Appleman. New York: Norton, 1976.

R. Ganeshan (✉) · T. Boone
Mason School of Business, College of William and Mary, 23185 Williamsburg, VA, USA
e-mail: Ram.ganeshan@mason.wm.edu

T. Boone
e-mail: Tonya.Boone@mason.wm.edu

V. Jayaraman
Department of Management, School of Business Administration,
University of Miami, 33124 Coral Gables, FL, USA
e-mail: vaidy@miami.edu

T. Boone et al. (eds.), *Sustainable Supply Chains*,
International Series in Operations Research & Management Science 174,
DOI 10.1007/978-1-4419-6105-1_1,

> The data now show that greenhouse gas emissions are accelerating much faster than we thought... Over the last decade developing countries such as China and India have increased their electric power generation by burning more coal. Economies in the developing world are becoming more, not less carbon-intensive. We are definitely in unexplored terrain with the trajectory of climate change, in the region with forcing, and very likely impacts, much worse than predicted in the fourth assessment [of the IPCC][2]... One thing that seems to be certain, however, is that as a society we are facing a climate crisis that is larger and harder to deal with than any of us thought. The sooner we take decisive action, the better our chances are of leaving a sustainable world to future generations.

The crisis, a possible 1.8–4°C increase in global temperatures as a result of unchecked emissions resulting from human activities (including a multitude of large scale supply chains) can have disastrous effects. Scientists warn that human-caused climate change can cause large-scale species extinction; disrupt our food supply; change weather patterns and increase sea levels; cause large-scale migration and flood coastal areas; and finally, can stress public health systems due to spread of diseases. The costs will be shared by all of us, and the worst affected will be the ones in the most impoverished countries.

Technological innovations alone will not create a sustainable urban environment; the economic and political climate must be such that sustainable practices are proactively adopted by various stakeholders. This will require innovative action by policymakers to incorporate the costs of unsustainable negative externalities into the economics of our energy decisions. Greenhouse gas emission, air and water pollution and low-efficiency building practices are critical externalities in which the decision-maker does not incur the costs of unsustainable practices. In the current trajectory, individuals and businesses that use energy-intensive products and services do not pay for the cost of emitting greenhouse gases, encouraging unsustainable patterns of energy usage.

Sustainable development is a global challenge that calls for a coordinated response from all of the world's communities. Countries, cities and local governments need to measure, manage, and mitigate emissions and enact regulations that support sustainable development. Multiple treaties and initiatives are already underway and have made significant progress. The United Nations' Kyoto protocol, for example, curbs emissions by developed countries. The 2009 Copenhagen talks, while considered a failure, has prompted countries—both developed and developing—to set voluntary emission targets. It is undeniable that the future populations shall be highly urbanized. Megacities of the future are the only way in which the productivity of massive populations can be fully realized. The proportion of the global population in urban environments compared to those in rural areas in 1950 was 29% urban to 71% rural. In 2010, the figure has changed to 51% urban to 49% rural. Projected figures of 2050 indicate a ratio of 70% urban to 30% rural populations[3]. Presently we use around 495 quadrillion Btu of energy to power our lives. Out of this, liquid fuels and natural gas supply add up to 58%, 26% by burning coal, 10% from renewable and 5.5%

---

[2] http://carnegiescience.edu/news/decisive_action_needed_warming_predictions_worsen_says_carnegie_scientist.

[3] UN Population Statistics: esa/un.org/unup.

from nuclear sources. About a fifth of this total (97.94 quadrillion Btu of energy) is used for transportation services[4]. In response, the C40 group, a group of the world's largest cities, share information on how to promote sustainable development and many cities have set ambitious targets to curb emissions as part of this initiative[5].

Sustainable development has also had a profound impact on many businesses, their operations and their supply chains. The complex and interrelated nature of environmental problems will challenge firms to move beyond mere compliance with environmental regulations to develop innovative responses to environmental challenges.[6] Organizations will have to focus on innovative responses such as reducing waste and emissions, recycling solid waste, conserving energy and reducing business impacts on the environment. The National Academy of Sciences, an American body, predicts that energy use can be cut by 30% by 2030, simply by implementing energy-efficient technologies that already exist. However, the removal of total societal cost from the energy-use equation is currently preventing market forces from selecting the most sustainable way forward. Addressing this will require innovative policy from our senior administrators and policy makers, but will have an enormous impact.

In a quest to meet customer demands for more environmental responsible products, companies are rethinking supply chain strategies, structures and flows. Sustainability-related initiatives are motivating new forms of inter- and intra-firm collaboration. Meanwhile, new partnerships are requiring increasing levels of operational transparency and accountability.

At the same time, many companies are finding out that evaluating their supply chains through the lens of sustainability can provide a considerable competitive advantage. Sustainable supply chain management activities—which attempt to reduce the environmental impact of products throughout their entire life cycles, can motivate changes which lead to increased quality, improved efficiency and greater organizational effectiveness. In the future, it is those organizations that are able to secure the resources and proactively develop competencies to address the challenges of natural environmental constraints that will win the battle.

The intent of this edited volume is to provide a business case for sustainable operations through tutorials, thought pieces, and rigorous empirical and theoretical research. We have put together an eclectic group of chapters with the hope of bringing together a discussion of the salient factors that support a sustainable enterprise, as well as analyses of expected consequences of sustainability initiatives—both positive and negative.

The authors featured in this edited volume—both by academics and practitioners—explore leading-edge ideas to (re)design and market products, source, manufacture, and eventually distribute and recover or dispose them in an environmentally-, culturally-, and socially-responsible way.

---

[4] International Energy Outlook 2010 US Energy Information Administration- eia.doe.gov.

[5] See Chap. 2 on more details on country and city emissions and some of the established targets.

[6] Hart, S. A natural-resource based view of the firm. Academy of Management Review. 1995, 20(4), 986–1014.

1.2 A Brief Evolution of Sustainability

Jean Baptiste Joseph Fourier in 1824 provided the first reference to greenhouse gases when he proposed that the earth's atmosphere acted like a glass pane, trapping the sun's heat. Svante Arrhenius, in 1896, was the first to suggest that levels of carbon di-oxide in the atmosphere impacted global temperatures via the greenhouse effect.

The US marine biologist and conservationist Rachel Carson's 1962 book *Silent Spring* studied the harmful effects on pesticides such as DDT on animals, especially birds. Her thesis was that the wide spread use of pesticides was killing birds, leading to a "silent spring"—devoid of singing birds. The book was widely read and inspired a grassroots environmental movement, leading to the creation of organizations such as Greenpeace and regulatory bodies such as the Environmental Protection Agency. The book ultimately led to the ban of DDT as a pesticide in 1972.

The first major conference under the aegis of the United Nations—The Stockholm Conference on the Human Environment—was held in 1972. The conference was attended by representatives from 113 countries, 19 inter-governmental agencies, and more than 400 inter-governmental and non-governmental organizations with the intent of understanding environmental issues and sustainable development. It led to the creation of the United Nations Environmental Program (UNEP) and the establishment of environmental regulatory agencies by many of the participating governments. The conference proclaimed that:

> The protection and improvement of the human environment is a major issue which affects the well-being of peoples and economic development throughout the world; it is the urgent desire of the peoples of the whole world and the duty of all Governments.[7]

The then Prime Minister of India, Indira Gandhi, commented at the Stockholm conference:

> . . . the environment cannot be improved in conditions of poverty, unless we are in a position to provide employment and purchasing power for the daily necessities of the tribal people and those who live in around our jungles, we cannot prevent them from combing the forests for food and livelihood; from poaching and from despoiling the vegetation. How can we speak to those who live in villages and slums about keeping the oceans, the rivers and the air clean when their own lives are contaminated at the source?

Her comment linked economic and social conditions, the environment, and sustainable development—all of which have to be integrated for a viable public policy.

The Stockholm conference was just the start of inter-governmental talks on the environment. The so called "Earth Summit" held in Rio de Janeiro in 1992 produced the United Nations Framework Convention on Climate Change (UNFCC)—an international treaty whose goal is to "fight global warming and to cope with the inevitable consequences of human-caused climate change." Since the 1992 United Nations

[7] http://www.worldservice.org/stockholm.html.

Earth Summit, sustainable development (entails both economic growth and environmental sustainability) has been gaining ground among several corporate leaders. However, Robert Shapiro, CEO of Monsanto said it best:

> We can't expect the rest of the world to abandon their economic aspirations just so we can continue to enjoy clean air and water. This is neither ethically correct nor likely to be permitted by the billions of people in the developing world who expect the quality of their lives to improve.

The Kyoto Protocol, an addition to the UNFCC treaty sets legally binding measures for 37 industrialized countries. The Protocol's first commitment period begun in 2008 and ends in 2012. Talks are underway to extend the commitment period well beyond 2012.

The intergovernmental Panel for Climate Control (IPCC), established in 1988, is the leading international body for the assessment of climate change. It was established by the United Nations Environment Program (UNEP) and the World Meteorological Organization (WMO) to provide "the world with a clear scientific view on the current state of knowledge in climate change and its potential environmental and socio-economic impacts.[8]" The IPCC has so far published four assesments on climate—the fifth is underway and will be published in 2014. The IPCC was jointly awarded the Nobel prize in 2007 along with US Vice President Albert Arnold (Al) Gore Jr.[9] "for their efforts to build up and disseminate greater knowledge about man-made climate change, and to lay the foundations for the measures that are needed to counteract such change[10]." Figure 1.1 gives a timeline of the major events related to sustainable development and the environmental movement in the last 50 years[11].

Over the last two decades, firms are increasingly incorporating Corporate Social Responsibility (CSR) into their business models. The key idea is that, in addition to the profit motive, firms are obligated to a wide variety of stakeholders—their employees, customers, the environment, and to the betterment of society. Most major firms publish CSR reports and established guidelines make it easy to report their ideas, policies, and actions to their stakeholders.

For example, the Global reporting Initiative (GRI) established in 1998 sets guidelines to report on sustainability efforts. The GRI Framework can be used "to demonstrate organizational commitment to sustainable development, to compare organizational performance over time, and to measure organizational performance with respect to laws, norms, standards and voluntary initiatives[12]." Over 1,000 firms around the world use the GRI reporting guidelines to inform their stakeholders of their targets and their progress towards it.

---

[8] http://www.ipcc.ch/organization/organization.shtml.

[9] In the documentary *The Inconvinient Truth*, Vice President Al Gore takes his message of the environment in crisis to a global audience. The documentary is widely credited for reenergizing the environmental movement.

[10] http://www.nobelprize.org/nobel_prizes/peace/laureates/2007/.

[11] An interactive version of the timeline is available at: http://operationsbuzz.com/2010/03/sustainability-a-brief-timeline/.

[12] http://www.globalreporting.org/AboutGRI/WhatIsGRI/.

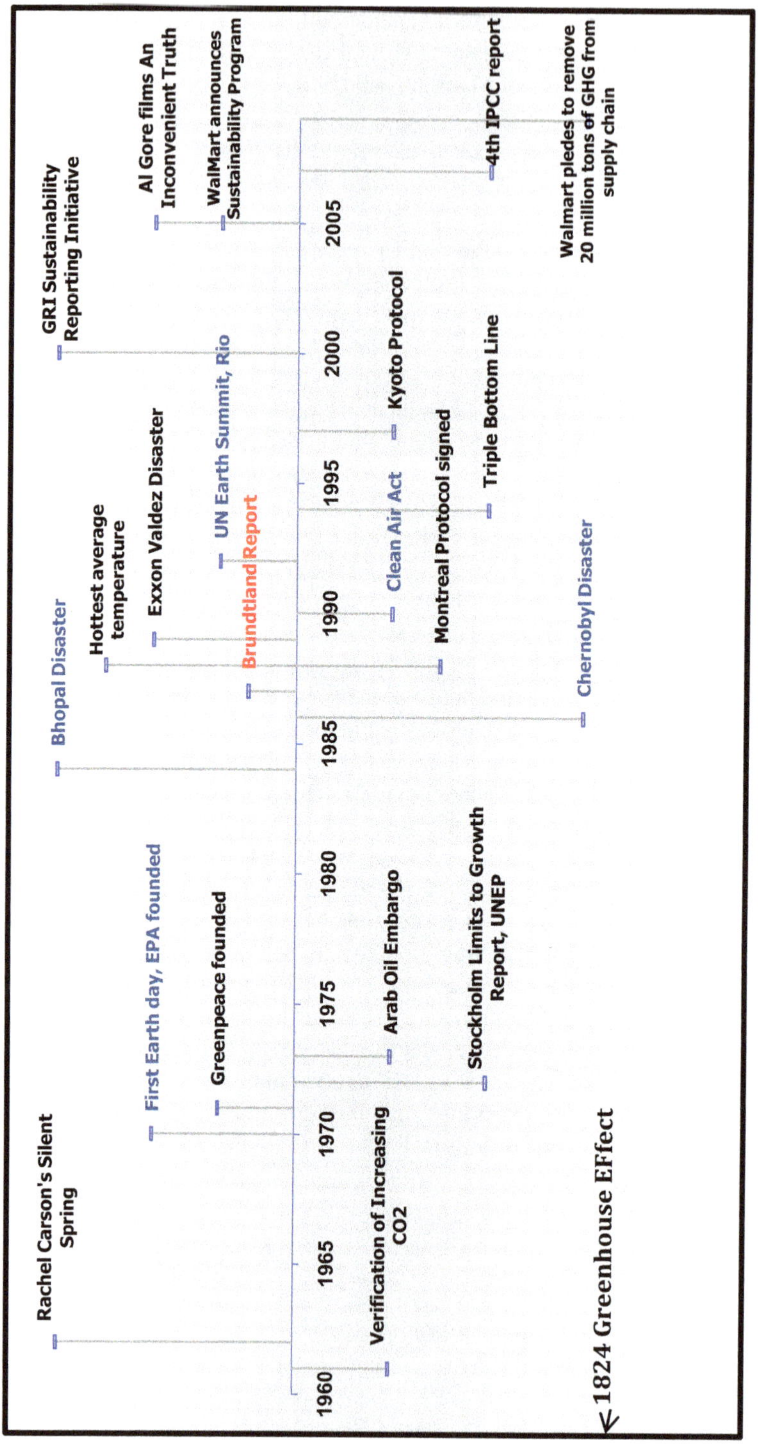

Fig. 1.1 A timeline for sustainability

Most firms focus on three major areas: (1) Environmentally-sensitive Design, (2) Efficiency, and (3) Recovery[13]. Firms are designing products that encompass the entire life cycle—from raw material sourcing to product use and recovery. Products and services are designed to use fewer inputs, smaller quantities of each input, and inputs that are recycled. Products are energy efficient and use less energy when in use and at end of life, can be dismantled, reclaimed, and reused.

Efficiency measures include using less energy and water in the supply chain, and reducing waste streams generated in procurement, production, and distribution. Finally, recovery involves taking back the product after its use with the intent of creating reusable raw materials. Multiple models have emerged to address sustainability in the firm—for example William McDonough and Michael Braungart's "Cradle to Cradle" philosophy[14] or a myriad of carbon abatement schemes—all of which provide guidelines for "closing the loop" in the supply chain.

The IPCC reports that to stabilize carbon dioxide levels between 445 and 490 parts per million (ppm)—resulting in an estimated global temperature increase of 2–2.4°C above the pre-industrial levels—emissions would need to peak around 2015, along with a 50–85% reductions on 2000 levels by 2050. While, the current trajectory of public policy and organizational response to sustainable development is encouraging, it falls well short of the reductions recommended to maintain CO_2 concentrations in the 445–490 ppm range leaving much of humankind susceptible to the risks of climate change.

1.3 How This Book is Organized

This edited volume is made up of 15 Chapters including this editorial. Each of the chapters are independent and can stand alone as a paper on the topic it is addressing. While there is no clear way to classify these papers, we have grouped them into loose clusters.

Chapter 2 shows the greenhouse emissions at various levels, from countries all the way to individual products. Chapters 3–7 each focus on an industrial sector and address issues specific to that industry. Chapter 3, for example, addresses the use of bio-fuels in Oil & Gas. Chapter 4, via case studies, discusses how healthcare firms are organizing for sustainability. Chapter 5 also focuses on healthcare, and describes and develops a sustainable network design/redesign model for the complex supply chain of human blood, which is a valuable yet highly perishable product. Chapter 6 makes a case, with a special emphasis on India, on how sustainability could be viewed as a key driver for profit and innovation in IT companies. And Chap. 7 presents a case study on LEED certification of Miller Hall, home of the Mason School of Business where two of the authors (Tonya and Ram) work.

[13] See Chap. 13 for examples.

[14] http://www.mcdonough.com/cradle_to_cradle.htm.

Chapters 8–10 address product take back in the supply chain. Chapter 8 introduces e-waste and surveys what firms are doing to combat it. Chapter 9 provides an overview of existing take-back legislation and academic papers that have studied various research questions associated with them. Chapter 10 is a tutorial that addresses the problem of product disposition on a closed-loop supply chain: what should a firm do with a product return?

Chapters 11–15 address measurement, monitoring, decision-making, and reporting regarding environmental issues in a firm. Chapter 11 provides an academic survey of eco-labeling and the consumer's willingness to pay for them. Chapter 12 discusses how firms can measure the total carbon Footprint in their supply chains and some of the strategies they can use to mitigate carbon emissions. Using the price of call options, Chap. 13 illustrates how managers can quantify the savings attributed to sustainability-related investment. This technique provides a single number that managers are more familiar with when allocating a fixed investment budget among competing project proposals. Chapter 14 develops a non-linear optimization model that addresses the complex trade-offs involved in making joint operational and environmental decisions. The model encompasses tradeoffs across product life-cycle stages—starting with product design, through production, to the end of the product's economic life. Finally, Chap. 15 develops a Data Envelopment Analysis-based method for supplier evaluation incorporating environmental and business factors. This technique allows the joint consideration of a number of performance measures as well as a tool for comparative analysis for suppliers.

Chapter 2
By the Numbers: A Visual Chronicle of Carbon Dioxide Emissions

Tonya Boone and Ram Ganeshan

Abstract Carbon dioxide equivalent (CO_2-eq.) emissions are now considered a de facto indicator of environmental impact. Through a series of visuals, this chapter highlights the size and scope of carbon emissions at multiple levels—countries, cities, industrial sectors and products.

2.1 Introduction

The fourth assessment of the Inter Governmental Panel on Climate change (IPCC) reports that:

> Warming of the climate system is unequivocal, as is now evident from observations of increases in global average air and ocean temperatures, widespread melting of snow and ice and rising global average sea level.[1]

Since the 1750s, atmospheric concentrations of carbon dioxide have risen from about 280 to 379 parts per million (ppm) in 2005. This increase is largely attributed to large-scale supply chains that sustain modern economies and lifestyles. The biggest contributor to greenhouse gases (GHG) emissions are energy use to run industrial processes, generate electricity, transport goods, and heat and cool residential and industrial structures. Changes in land use, agricultural and industrial processes, and waste management are the other major contributors to GHG in the environment.

GHG emissions has led to a 0.6°C increase in the global average surface temperature since 1900. If the current trends in emissions continue, the IPCC estimates that global temperatures will rise a further 1.4–5.8°C by 2100. Scientists agree that such

All web sites referenced in this Chapter were accessed 15th August 2010.

[1] IPCC Synthesis Report, p. 30, http://www.ipcc.ch/publications_and_data/publications_and_data_reports.shtml.

R. Ganeshan (✉) · T. Boone
Mason School of Business, College of William and Mary, 23185 Williamsburg, VA, USA
e-mail: Ram.ganeshan@mason.wm.edu

T. Boone
e-mail: Tonya.Boone@mason.wm.edu

T. Boone et al. (eds.), *Sustainable Supply Chains*,
International Series in Operations Research & Management Science 174,
DOI 10.1007/978-1-4419-6105-1_2, © Springer Science+Business Media New York 2012

increases will likely disrupt eco-systems (about 30% of plant and animal species will face extinction); increase environmental and climate disasters; disrupt food supply, and cause widespread health issues. Such disruptions will impact supply chains in every country and in every industry and the consequences are likely to be detrimental to a large portion of the world's population, especially in the world's impoverished countries.

There is much agreement today that to keep the global average temperature from increasing more than 2°C above pre-industrial levels, GHG emissions would need to peak around 2015 and subsequently decline by 40–45% by 2050 compared to 1990 levels. This will involve a concerted efforts at all levels—countries, cities, local governments, corporations, and individuals. This effort needs to be focused on how we generate and use energy, increase the efficiency—both ecological and economic—of our industrial and agricultural processes, and finally an effort to "close the loop"—to conserve and reuse natural resources in supply chains.

Greenhouse gas emissions are often reported in terms of "carbon dioxide equivalents" (CO_2-eq.). Emissions from other gases such as methane, nitrous oxide, hydrofluorocarbons, perfluorocarbons, and sulfur hexafluoride are translated into their CO_2 equivalent emissions using what is called the global warming potential (GWP) over a specified time frame, usually 100 years. CO_2 has a GWP of 1 since it is the base unit against which all other GHGs are measured. Methane, for example, has a GWP of 25 which means that 1 ton of Methane has the same global warming potential as 25 tons of CO_2 over a 100-year horizon. For better or worse, GHG emissions in CO_2-eq. is increasingly a benchmark on how to measure, manage, and mitigate GHG emissions. International bodies such as the United Nations Framework Convention on Climate Change (UNFCCC) use CO_2-eq. to track national emission trends and inform the global community on human-induced interference with the climate system. National and local government leaders use GHG emissions in CO_2-eq. to develop policies for emissions reductions and to track the progress of those policies. Regulatory bodies rely on CO_2-eq. measurements to establish compliance records with allocated emission rates. Businesses use CO_2-eq. to better access the risks and opportunities of climate change, and finally, individuals use CO_2-eq. measurements to make environmentally friendly life style choices.

Our intent in this Chapter is to provide a snapshot of CO_2-eq. emissions from different entities—countries, cities, industrial sectors, an average firm, and individual products. Our goal is to inform the reader on the size and scope of GHG emissions in CO_2-eq. from these entities—not to explain *why* the emissions are the size they are or prescribe *if* and *how* the emissions can be curbed. The text in this Chapter merely lays the context for the visuals.

While comparisons of CO_2-eq. emissions between countries, cities, industrial sectors, or products are inevitable, we offer a few caveats. First, while rigorous protocols are available to compute total CO_2-eq. quantities, they are not consistently applied, especially for smaller entities such as products and services. Without knowledge of exactly what is included in the final, often self-reported, CO_2-eq. number, even a comparison between any two similar entities may be misleading. Second, while CO_2-eq. is an excellent surrogate for an entity's environmental impact, it is not all encompassing. Without knowledge of the entity's impact on water, natural resources, eco-systems, and public health, effective policies and strategies cannot be formulated.

2.2 Country Emissions

Human caused climate change is a global problem and it is no surprise that tackling it is addressed through international treaties such as the Kyoto Protocol[2]. Figure 2.1a, b show the 2007 gross CO_2-eq. emissions and the per-capita emissions from the countries of the world. The developed economies of the USA, European Union, Canada, Japan, the Russian Federation, and Australia; and the developing countries of China, India, Brazil, and are the largest emitters of greenhouses cases. But when compared on a per-capita basis, the emissions of the developed countries are much higher than the developing or the under-developed countries.

The Kyoto Protocol uses the principle of 'common but differentiated responsibility' to tackle GHG mitigation. It sets binding targets for 37 industrialized countries (also called "Annex I") for reducing GHG emissions and exempts developing economies from emission targets. The members of the European Union and other European countries[3] will reduce their GHG emission by 8% against 1990 levels over the five-year period 2008-2012. Over the same time period, Canada and Japan will reduce 6% over 1990 levels. The reductions will be through national measures and market-driven mechanisms[4] that were introduced as part of the Kyoto protocol. The USA has not ratified the Kyoto protocol. China, India, and Brazil—as developing economies—and are exempt from emission targets per the Kyoto protocol.

Figure 2.2a shows the trends in CO_2-eq. emissions from Annex I countries from 1990 through 2009. Figure 2.2b gives the percentage growth or reduction in CO_2-eq. relative to 1990, the base year per the Kyoto Protocol. The total Annex I emissions are down 11.3% compared to the 1990 base year. In the same time period, the EU-27 reduced emissions by 17.4% while the US increased its emissions by 7.2%. Meanwhile, China's emissions have grown 189.5% and India's by 126.1% from 1990 through 2007.

Efforts to legally extend carbon mitigation beyond the Kyoto Protocol are currently stalled—however the delegations attending the failed Copenhagen Summit have agreed "to take note of" but not "adopt" what is known as the Copenhagen Accord[5]. The Copenhagen Accord was drafted by the US and developing economies of China, India, Brazil, and South Africa (the BASIC countries). The Accord endorsed "the scientific view that the increase in global temperature should be below 2°C" and that Annex I countries will "commit to economy-wide emissions targets for 2020."

---

[2] The Kyoto Protocol was signed on 11th December 1997 in Kyoto and went into force on 16th February 2005.

[3] This includes the 15 members in 1997 plus Bulgaria, Czech Republic, Estonia, Latvia, Liechtenstein, Lithuania, Monaco, Romania, Slovakia, Slovenia, and Switzerland. The members of the EU 15 will use this as a joint target, i.e., the total amounts to 8% reduction while individual countries differ in their reductions. Hungary and Poland have a 6% reduction target.

[4] http://unfccc.int/kyoto_protocol/mechanisms/items/1673.php.

[5] http://unfccc.int/meetings/cop_15/copenhagen_accord/items/5262.php.

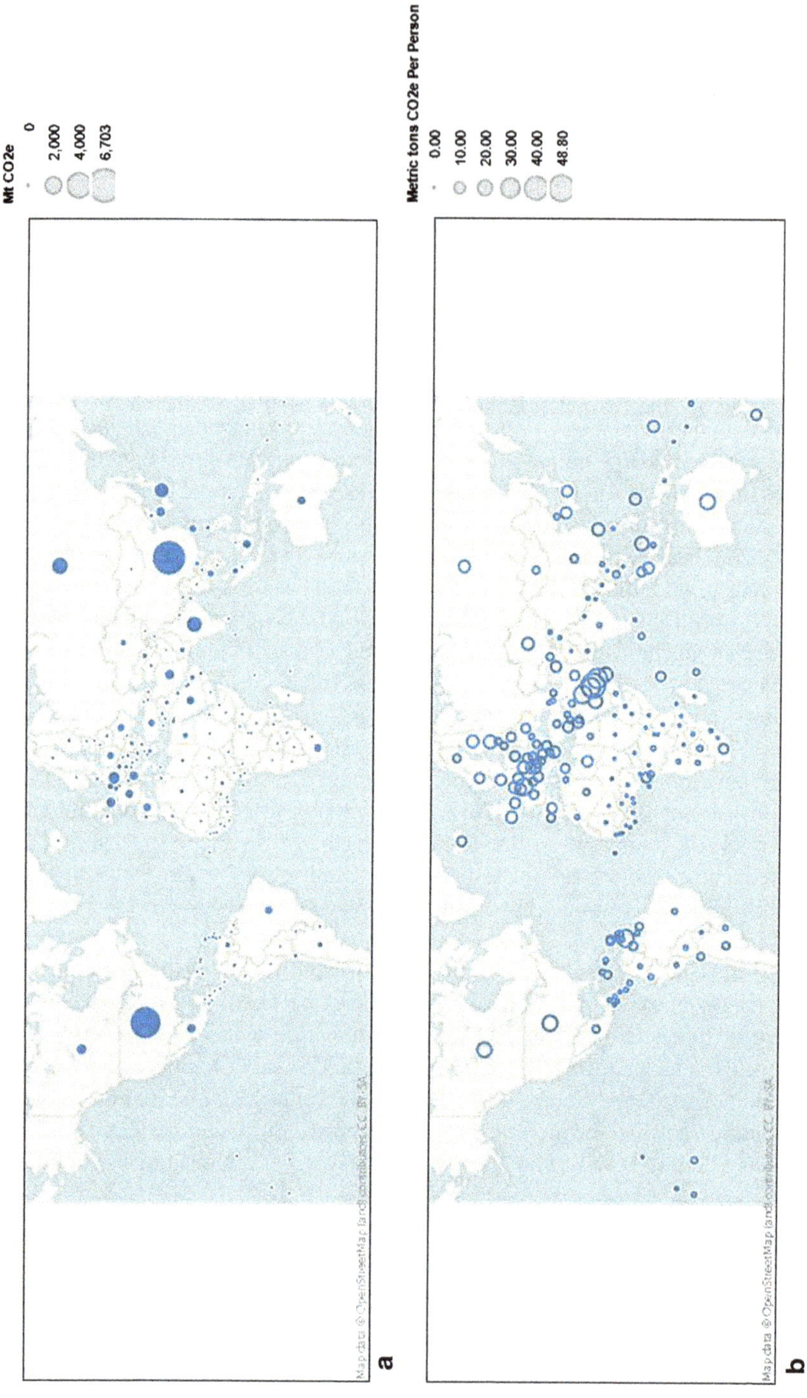

Fig. 2.1 **a** 2007 global Co_2-eq. Emissions (Kilotons). **b** 2007 global per capita Co_2-eq (Mt/person)

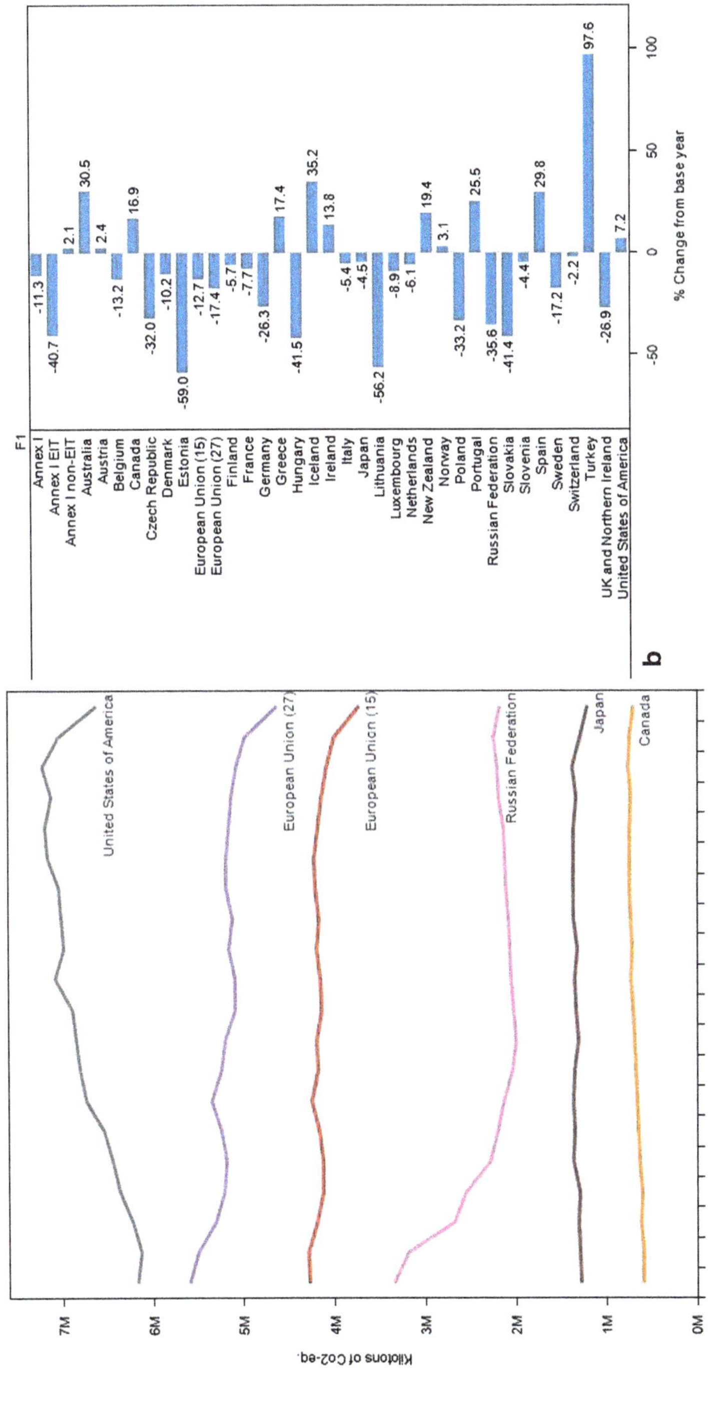

Fig. 2.2 **a** Trends in GHG Emissions (Annex I). **b** %change in GHG Emissions in 2009 compared to 1990

The Accord also recognized that developing nations would have to "implement mitigation actions," report their emissions, and subject it to verification.

Although not legally binding, several countries have proposed reduction targets in connection with the Copenhagen Accord. Compared to 1990 levels, the EU has proposed a target reduction of 20–30%, Japan by 25%, and Russia by 15–25%. The US and Canada have proposed a 17% cut by 2020 compared to 2005 levels. China has proposed a cut of 40–45%; and India a cut of 20–25% on emission intensity[6] by 2020 compared to 2005 levels[7].

2.3 City Emissions

In 2010, about half of the world's population lived in cities. Cities consume about 75% of the world's energy and are responsible for anywhere from 40–70% of the world's GHG emissions. Figure 2.3 gives the total and the per-capita GHG emissions for selected cities[8]. For example, Tokyo has emissions of 65.9 million metric tons with a per-capita of 5.16 metric tons/person.

Cities, as the hub of social and economic activity, are both significant contributors GHG emissions and highly vulnerable to climate change. City governments oversee infrastructure, host and promote businesses, and provide services for their citizens. Carefully assessing climate risk and taking steps to reduce their exposure to climate change will create a better business environment and improve the quality of life of its citizens.

Cities have a major role to play in GHG mitigation. First, city governments and services can have a substantial carbon footprint. For example, New York City government operations account for 3.47 million metric tons of CO_2-eq. annually. Tokyo's city government emissions are 2.06 million metric tons annually[9]. Using renewable energy, improving efficiency and expanding public transport, retrofitting public buildings and street lighting, managing waste streams, and increasing green spaces all bring down the footprint and enable a better quality of life. Second, cities have a have a significant impact on how their citizens live and how corporations conduct business. Engaging citizens and businesses through incentives and joint partnerships will galvanize cities towards being climate neutral.

Cities are already taking major steps to battle climate change. For example, a group of 40 large cities have banded together to form the C40 city group. Partnering with the Clinton Climate Initiative (CCI), they plan to reduce energy use and

---

[6] Emission intensity is the GHG emissions per unit of GDP. This unit does not ensure a reduction in absolute emissions—only that emissions will grow slower than GDP growth.

[7] Targets established by countries in the Copenhagen Accord are available at: http://unfccc.int/meetings/cop_15/copenhagen_accord/items/5262.php.

[8] Data is from cdpproject.net. A group of the world's largest 40 cities (c40cities.org) in collaboration with the Clinton Climate Initiative report their GHG emissions through the CDP project. The absolute GHG emissions are self-reported. Where the per-capita is not reported, it was extrapolated by dividing the absolute GHG emissions by the population of the city reported at C40cities.org.

[9] City responses to CDP questionnaire.

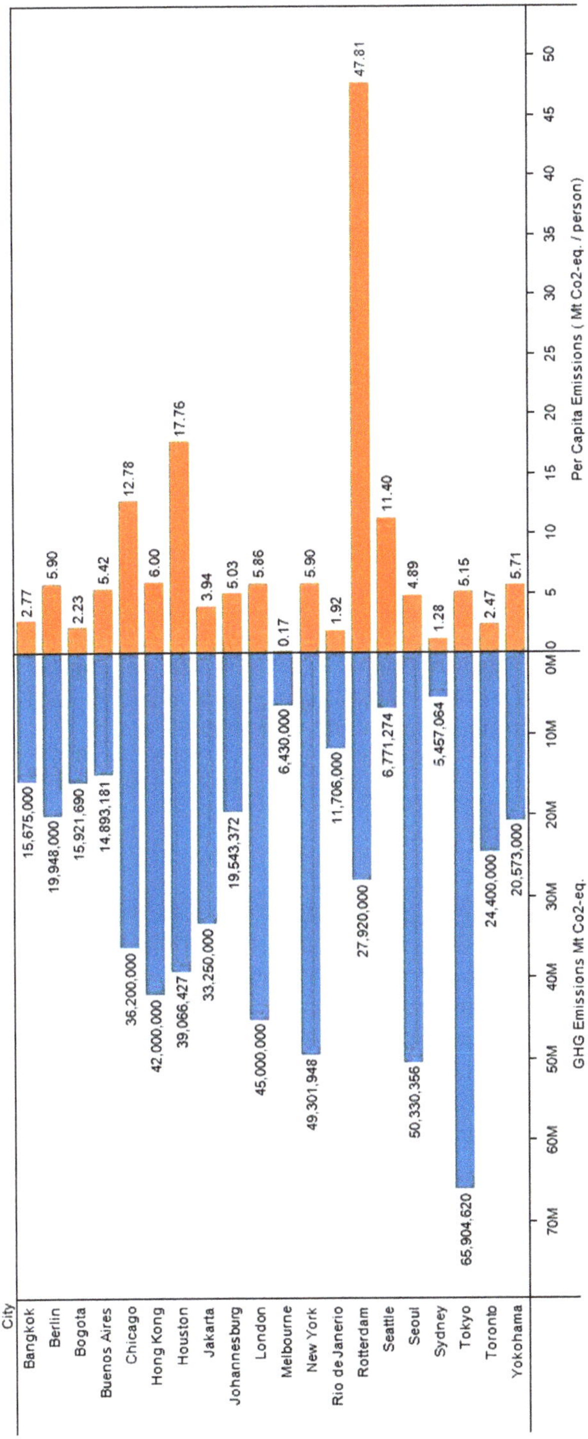

Fig. 2.3 Emissions of major cities

greenhouse gas emissions, by providing "direct assistance to individual cities, creating a purchasing consortium to pool the purchasing power of these cities, and facilitating the sharing of information about successful and replicable programs[10]." Programs involve building retrofitting, low-carbon transportation, waste management, and climate-positive growth. The C40 reports their emissions and mitigation strategies via the Carbon Disclosure Project's (CDP) Project Cities program.

ICLEI (Local Governments for Sustainability), an association of over 1220 local governments, also has several programs that help cities collaboratively combat GHG emissions. ICLEI's goals are to help local governments plan programs and policies for sustainable development. Among others, this includes creating baselines, setting GHG emission targets, meeting these targets through well-defined projects, and monitoring, reporting, and verifying GHG mitigation strategies.

One of the programs to which the ICLEI provides technical and strategic assistance is the World Mayors Council on Climate (WMCC), a group of Mayors of major cities committed to mitigating the impact of climate change. Over 191 Mayors and local authority representatives have signed the "Mexico City Pact[11]"—a pledge to voluntarily reduce GHG emissions, set targets, take GHG mitigating actions, and report them in a transparent manner.

Through efficiency and engagement programs cities are setting aggressive targets for emissions. For example Portland and Seattle, USA; Toronto, Canada; and Yokohama, Japan have targeted an 80% decrease in CO_2-eq. emissions by 2050 over 1990 levels. Rotterdam has pledged a cut in emissions by 50% and London has proposed a cut of 60% in emissions by 2025 over 1990 levels.

2.4 Sector Emissions

Businesses are responsible for 20–25% of the world's GHG emissions. Emissions for a firm are a result of their production processes and facilities, resources they consume, extraction and processing of relevant raw materials in supplier operations, and the impact of their product or service on their customers.

Firms are increasingly accessing the risk and opportunities of climate change and integrating it into the overall business strategy. In the latest survey by the CDP of the top 500 global firms[12], 48% said that they integrate risks and opportunities of climate change into their planning and 65% of them indicated that they had GHG emission mitigation targets. The major reasons for reducing GHG emissions include the increased ability to market new low-carbon products, compliance to existing and anticipated laws, and to increase efficiency while reducing disruptions in the supply chain.

Figure 2.4 gives the emissions from an "average firm" by sector. The data was collated from the Carbon Disclosure Project's "Investor CDP" program. Figure 2.5 also

[10] http://www.c40cities.org/about/aboutclinton.jsp.

[11] http://www.worldmayorscouncil.org/mexico-city-pact/read-the-pact.html.

[12] 410 firms responded to the survey. See cdpproject.net.

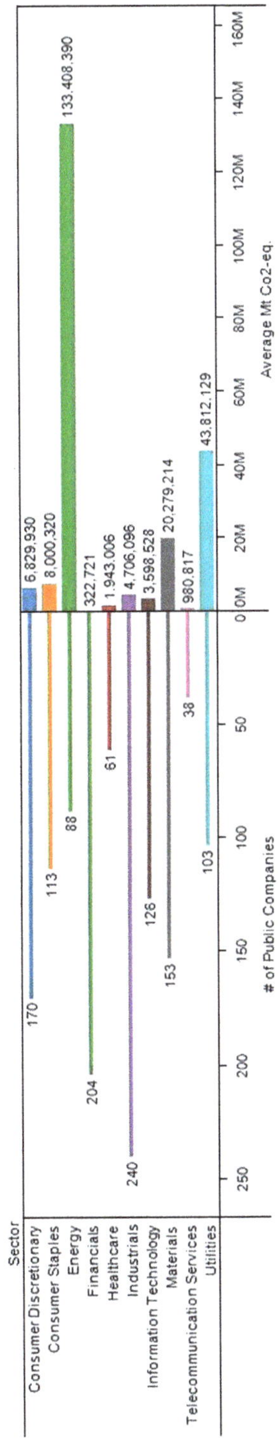

Fig. 2.4 Emissions of average firm by sector

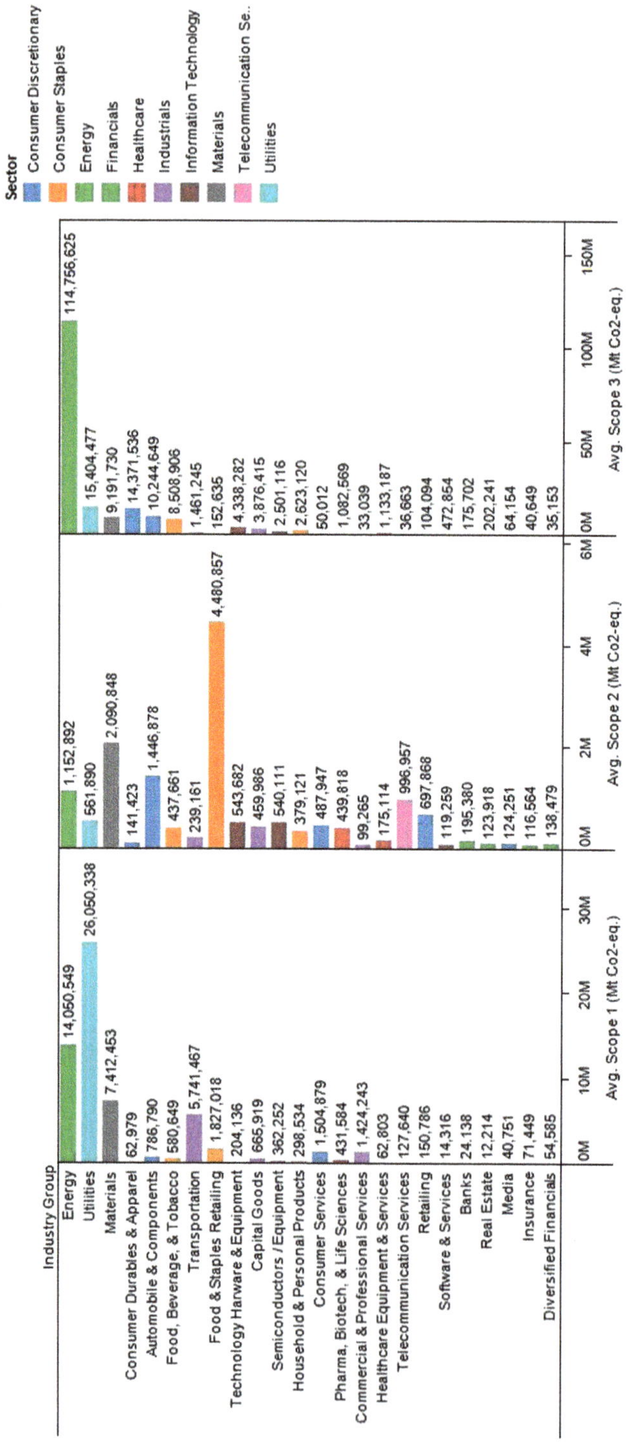

Fig. 2.5 Scopes by sector

shows the number of firms included to compute the average for the sector. For example, a sample of 88 firms in the Energy sector yielded an average of 133,408,390 Mt of CO_2-eq. while 204 firms in the Financial sector had an average emission of 322,721 Mt CO_2-eq.

There are multiple standards available to companies for accounting their GHG emissions. The GHG Protocol, developed in partnership with the World Resources Institute (WRI) and the World Business Council for Sustainable Development (WBCSD), is the most widely used standard. It provides businesses a consistent, and verifiable way to account for their GHG emissions. The GHG Protocol specifies three types of emissions called "scopes." Scope 1 emissions are a result of all company operations—production processes, waste streams, "fugitive" emissions[13], facilities, and employees. Scope 2 is emissions from purchased electricity, and Scope 3 is emissions associated with the supply chain. This includes related emissions from suppliers who handle product components; and the emissions from downstream distribution, use, and end-of-life management of the product/service[14].

Figure 2.5 gives the breakdown of the three scopes by industry group[15]. The scope information gives an insight into appropriate strategies to reduce emissions. Scope 3, not surprisingly, for the Energy sector is almost an order of magnitude larger than the sum of scope 1 and scope 2. While Oil and Gas exploration, drilling, refining, transportation, and storage are responsible for large GHG emissions, the use of oil and gas by other industrial sectors, residences, automobiles, trains, and planes are responsible for most of the emissions in this sector. The solution to reduce emissions in this industry is obviously complex—a national energy policy, fuel efficiency standards, urban planning and transportation all have important roles to play in addition to the firms in the sector. Figure 2.6 also shows that for manufacturing-intensive groups, scope 3 emissions are much higher than scope 1 or 2 suggesting that engaging supply chain partners is essential in GHG emission mitigation. The service-based industry groups have relatively lower emissions but on average have a higher scope 2 emissions compared to their scope 1 or 3, indicating that energy efficiency and a move to renewable fuels as a strategy for GHG mitigation.

2.4.1 Firm Snapshot: Apple Inc.[16]

Apple Inc. designs, manufactures and markets a wide range of electronic devices (personal computers, mobile communication and media devices, digital music

[13] These are simply leaks from the system.

[14] See the Chapter on "Carbon Footprinting: A Supply Chain Approach" for more details on the process.

[15] Firms typically report Scope 1 and 2 emissions. Since they have no direct control over Scope 3, many firms simply do not compute or report it. The averages in Fig. 2.6 is over the firms that have reported scope 3 information. So the sum of the scopes in Fig. 2.6 may not add up to the average reported emissions in Fig. 2.5.

[16] We have chosen Apple simply to illustrate GHG emissions in a sample firm. The choice was based primarily on Apple's strong brand recognition.

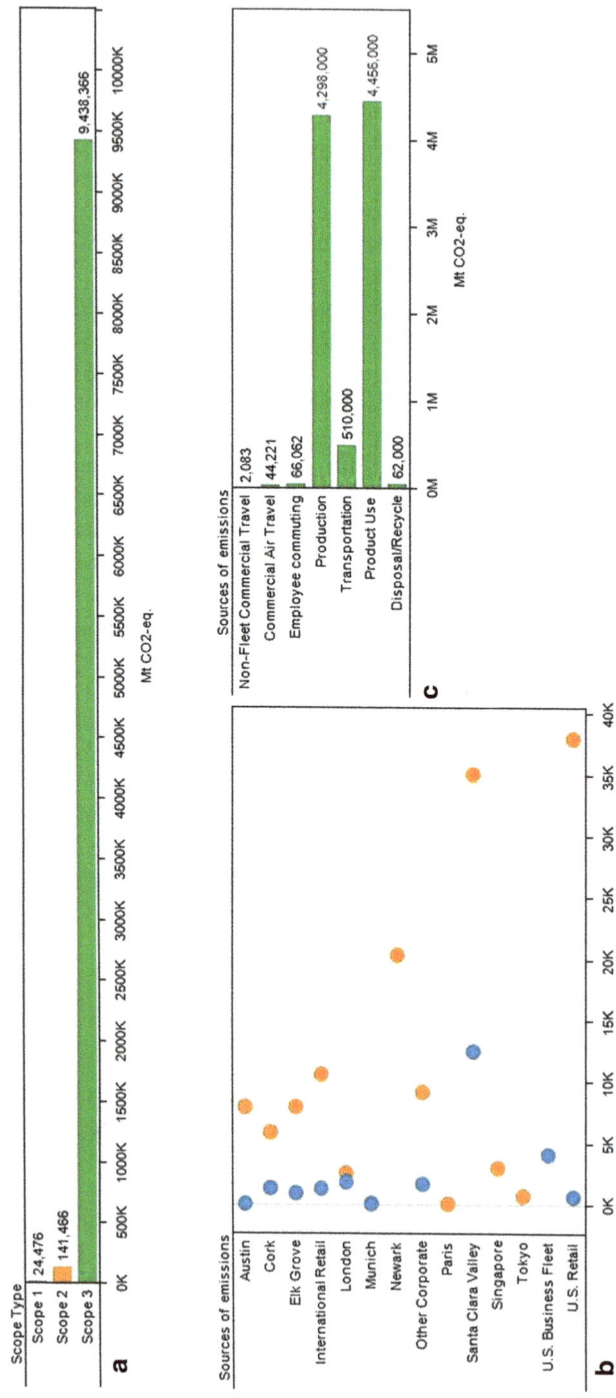

Fig. 2.6 **a** Apple GHG emissions by scope. **b** Scope 1 and 2 Breakdown by Facility. **c** Scope 3 Breakdown

players) software (OS X, iLife, etc.), services (iCloud), and third-party digital content and applications via its iTunes and App stores.

To manufacture its electronics devices, Apple sources components from all over the world, subcontracts production and assembly primarily in Asia, and sells in its own retail outlets in addition to several over distribution channels. Figure 2.6 gives the scope 1, 2, and 3 breakdown for Apple. Scope1 and 2 are primarily from its facilities and the breakdown between facilities is also shown in the figure. Scope 3 emissions, meanwhile, is more than 50 times the sum of the scope 1 and 2 combined. This includes among others, as Fig. 2.6 shows, the emissions from supplier operations, production, transportation, product use, and recycling. The two biggest sources of emissions are production and product use.

Apple's GHG reduction[17] efforts in production includes reducing material use in devices and in packaging, elimination of certain hazardous materials, and designing the product with appropriate materials such as Aluminum and glass so they can be recycled at end of life. The design also involves energy efficient components and it's monitoring via software so the emissions during product use are reduced. Apple has increased its reliance on renewable energy by 200% since 2005 and is using 100% renewable energy in its Elk Grove, Cork, and Austin facilities, reducing its Scope 1 and 2 emissions.

2.5 Product Emissions

In its most complete form, the CO_2-eq. emissions of a product includes activities throughout its life cycle right from extraction and processing of raw materials, production, use, and end-of-life. Figure 2.7a–e shows the CO_2-eq. of some common products and services[18].

Reporting the carbon emissions for a product or service, especially as a label on product packaging, is a recent phenomenon. Starting in 2007, Tesco, Britain's biggest retailer in conjunction with Carbon Trust (the UK-based developer of the PAS 2050 GHG standard) has carbon labels on more than 100 of its branded products such as pasta, potatoes, sugar, milk, laundry detergent, and orange juice. South Korea's environment ministry has introduced a "CooL Label," that covers products in multiple categories such as transportation and consumer electronics. E. Leclerc, a French retailer, has estimated the carbon emissions of 20,000 of its products and has launched an initiative where they can compute the total emissions of a consumer shopping cart[19]. Labeling initiatives are also underway in Thailand, Japan, Switzerland (Climatop label), Germany, and the USA.

While they help market products and inform consumers of the environmental impact, it is not clear if customers use the CO_2-eq. number on the label to guide purchase decisions. Second, standards for computing emissions for products are

---

[17] See http://www.apple.com/environment.

[18] Data sources are reported in the Appendix.

[19] http://www.jeconomisemaplanete.fr/.

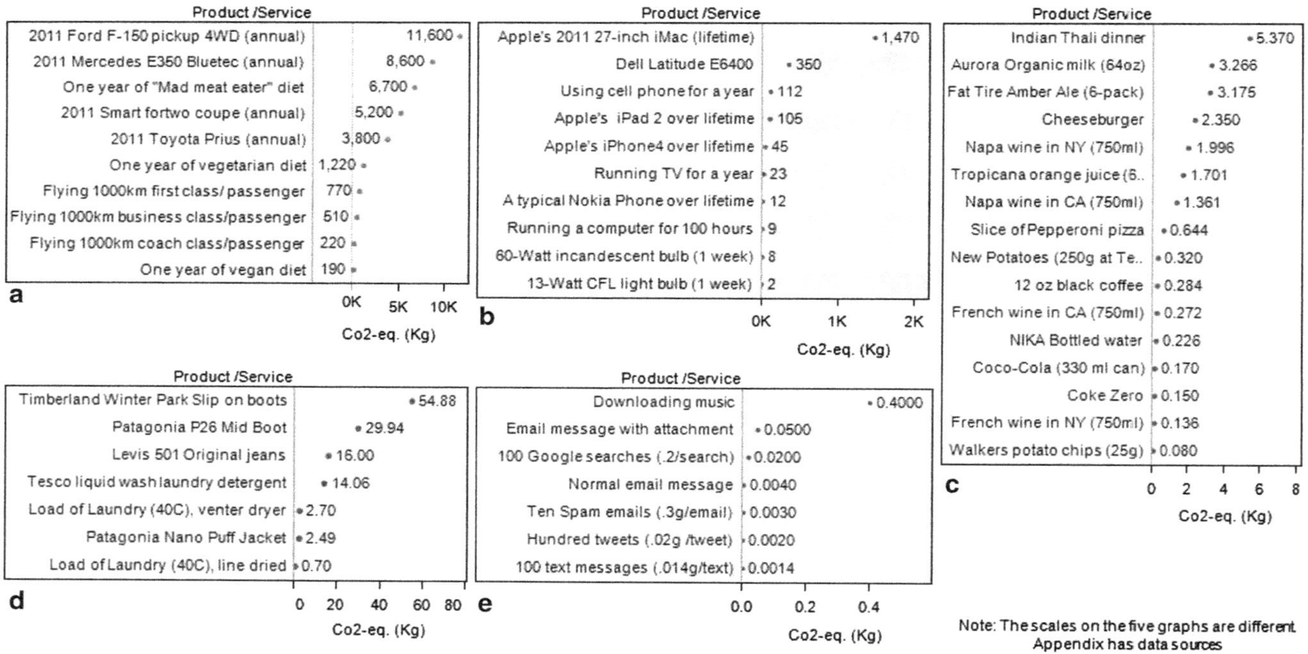

Fig. 2.7 **a** Travel and lifestyle, **b** Consumer electronics. **c** Food. **d** Clothing and washing. **e** Our digital social life

still evolving, and without a clear and consistent set of rules to compute the total emissions, customers can find the labels confusing.

While customer engagement through labeling and education is one way to mitigate GHG emissions, the emission analysis can have a significant impact on the firm reporting it.

The typical process is to map the supply chain, identify key activities in the product life cycle, and compute emissions. This analysis will identify carbon "hotspots" in the supply chain and appropriate actions can be taken to reduce emissions in those activities. This would typically involve engaging suppliers and customers to redesign products that have a smaller GHG impact, increasing efficiencies of processes, use of renewable energy, and reclaiming product after use to create raw material.

2.6 Conclusions

Our intent in this chapter was to give a snapshot of CO_2-eq. emissions of different entities. GHG mitigation will take actions at different levels—international agreements between countries, national measures, city and local government policies, firm level strategies, and finally individual choices—all aimed at reducing our impact on the environment. The policy issues or strategic considerations are multi-faceted and often involve multiple parties with differing objectives. However, we remain hopeful that entities at all levels will implement the required policies and execute the appropriate strategies to achieve the IPCC recommended 25–40% cut in emissions by 2050.

2.7 Appendix: Data Sources for the Visuals

This Appendix lists all data sources used to construct the visuals. All are available online.

2.7.1 Figure 2.1

GHG total and per-capita emissions: Climate Analysis Indicators Tool (CAIT) Version 8.0. (Washington:World Resources Institute 2011). Can be freely downloaded from: http://cait.wri.org.

2.7.2 Figure 2.2

Data is from the United Nations Framework Convention on Climate Change (UNFCCC) Secretariat. Parties to the convention are required per article 4 and 12 of the Convention to submit GHG inventories to the Secretariat. Can be freely downloaded from: http://unfccc.int/ghg_data/ghg_data_unfccc/time_series_annex_i/items/3814.php.

2.7.3 Figure 2.3

The Carbon Disclosure Project in a non-profit organization based in London, UK.Working with multiple partners, they have developed a reporting protocol that many organizations, including the c40 cities and the Clinton Climate Initiative use. The data for this figure was from the "CDP Cities" program. Responses to questionnaires sent to cities on climate initiatives and GHG emissions are available free, but registration on the website is required: https://www.cdproject.net/en-US/Programmes/Pages/CDP-Cities-2011.aspx.

2.7.4 Figure 2.4

Data was obtained from CDP's "Investor CDP" program for the year 2010: https://www.cdproject.net/en-US/Programmes/Pages/CDP-Investors.aspx.

2.7.5 Figure 2.5

Datawas obtained from CDP's "Investor CDP" program for the year 2010: https://www.cdproject.net/en-US/Programmes/Pages/CDPInvestors.aspx.

2.7.6 Figure 2.6

Apple Inc. Emissions data was compiled from Apple's 2010 response to CDP's "Investor CDP" program. Much of the data is readily available from Apple's web site: http://www.apple.com/environment.

2.7.7 Figure 2.7

Data was compiled from a variety of sources. Where possible, the firm that did the footprinting exercise is referenced.

Category	Product/Service	Source
Clothing and laundry	Timberland Winter Park slip on boots	http://online.wsj.com/public/resources/documents/FOOTPRINT.pdf
	Patagonia P26 mid boot	http://www.patagonia.com/us/patagonia.go?assetid=23430
	Levis 501 original jeans	http://www.levistrauss.com/sites/default/files/librarydocument/2011/1/e-valuate-web-content-012011-finalv3.pdf
	Tesco liquid wash laundry detergent	http://online.wsj.com/public/resources/documents/FOOTPRINT.pdf
	Load of Laundry (40 C), venter dryer	http://www.guardian.co.uk/environment/green-living-blog/2010/oct/21/carbon-footprint-email
	Patagonia Nano puff jacket	http://www.patagonia.com/us/patagonia.go?assetid=23429
	Load of Laundry (40 C), line dried	http://www.guardian.co.uk/environment/green-living-blog/2010/oct/21/carbon-footprint-email
Consumer electronics	Apple's 2011 27 in iMac (lifetime)	apple.com/environment
	Dell Latitude E6400	http://i.dell.com/sites/content/corporate/corp-comm/en/Documents/dell-laptop-carbon-footprint-whitepaper.pdf
	Using cell phone for a year	http://www.unep.org/publications/ebooks/kick-the-habit/
	Apple's iPad 2 over lifetime	apple.com/environment
	Apple's iPhone4 over lifetime	apple.com/environment
	Running TV for a year	http://www.unep.org/publications/ebooks/kick-the-habit/
	A typical Nokia phone over lifetime	http://www.nokia.com/environment/devices-and-services/creating-our-products/environmental-impact
	Running a computer for 100 h	http://www.unep.org/publications/ebooks/kick-the-habit/
	60 W incandescent bulb (1 week)	apple.com/environment
	13 W CFL light bulb (1 week)	apple.com/environment
Food	Indian Thali dinner	http://www.eatlowcarbon.org/#
	Aurora organic milk (64 oz)	http://online.wsj.com/public/resources/documents/FOOTPRINT.pdf
	Fat tire amber ale (6-pack)	http://online.wsj.com/public/resources/documents/FOOTPRINT.pdf
	Cheeseburger	http://openthefuture.com/cheeseburger_CF.html
	Napa wine in NY (750 ml)	http://www.drvino.com/2009/04/14/the-carbon-footprint-of-wine-in-national-geographic/
	Tropicana orange juice (64 oz)	http://www.nytimes.com/2009/01/22/business/22pepsi.html?ref=business
	Napa wine in CA (750 ml)	http://www.drvino.com/2009/04/14/the-carbon-footprint-of-wine-in-national-geographic/

Category	Product/Service	Source
	Slice of pepperoni pizza	http://www.eatlowcarbon.org/#
	New potatoes (250 g at Tesco, UK)	http://www.footprintexpert.com/registry/Pages/default.aspx
	12 oz black coffee	http://www.eatlowcarbon.org/#
	French wine in CA (750 ml)	http://www.drvino.com/2009/04/14/the-carbon-footprint-of-wine-in-national-geographic/
	NIKA bottled water	https://www.nikawater.org/eco-policy/
	Coco-cola (330 ml can)	http://www.coca-cola.co.uk/press-centre/2009/march/coca_cola_announces_the_carbon_footprints_of_some_of_its_best_loved_brands.html
	Coke zero	http://www.coca-cola.co.uk/press-centre/2009/march/coca_cola_announces_the_carbon_footprints_of_some_of_its_best_loved_brands.html
	French wine in NY (750 ml)	http://www.drvino.com/2009/04/14/the-carbon-footprint-of-wine-in-national-geographic/
	Walkers potato chips (25 g)	http://www.walkerscarbonfootprint.co.uk/walkers_carbon_footprint.html
Our digital social life	Downloading music	http://download.intel.com/pressroom/pdf/CDsvsdownloadsrelease.pdf
	Email message with attachment	http://www.guardian.co.uk/environment/green-living-blog/2010/oct/21/carbon-footprint-email
	100 Google searches (0.2/search)	http://googleblog.blogspot.com/2009/01/powering-google-search.html
	Normal email message	http://www.guardian.co.uk/environment/green-living-blog/2010/oct/21/carbon-footprint-email
	Ten spam emails (0.3 g/email)	http://www.guardian.co.uk/environment/green-living-blog/2010/oct/21/carbon-footprint-email
	Hundred tweets (0.02 g/tweet)	http://mehack.com/from-chirp-energy-tweet-100-j-something-tweet
	100 text messages (0.014 g/text)	How Bad are Bananas bad: The Carbon Footprint of Everything by Mike Berners-Lee, Profile Books, 2010. ISBN-10: 1846688914
Travel and lifestyle	2011 Ford F-150 pickup 4WD (annual)	http://www.fueleconomy.gov/feg/findacar.htm
	2011 Mercedes E350 Bluetec (annual)	http://www.fueleconomy.gov/feg/findacar.htm
	One year of "Mad meat eater" diet	http://www.unep.org/publications/ebooks/kick-the-habit/
	2011 Smart for two coupe (annual)	http://www.fueleconomy.gov/feg/findacar.htm
	2011 Toyota prius (annual)	http://www.fueleconomy.gov/feg/findacar.htm

Category	Product/Service	Source
	One year of vegetarian diet	http://www.unep.org/publications/ebooks/kick-the-habit/
	Disposable diapers average child uses (8/day/2 years)	http://www.epa.gov/oms/climate/420f05001.htm
	Flying 1,000 km first class/passenger	http://www.unep.org/publications/ebooks/kick-the-habit/
	Reusable diapers average child uses (8/day/2 years)	http://www.epa.gov/oms/climate/420f05001.htm
	Flying 1,000 km business class/passenger	http://www.unep.org/publications/ebooks/kick-the-habit/
	Flying 1,000 km coach class/passenger	http://www.unep.org/publications/ebooks/kick-the-habit/
	One year of vegan diet	http://www.unep.org/publications/ebooks/kick-the-habit/

Chapter 3
Sustainable Supply Chains for Bio-Based Fuels and Chemicals

Joseph B. Powell, Paul M. Ayoub, Sachin Kapila, Alex P. R. Nevill and Grahame Y. Buss

Abstract Supply chain sustainability must be considered in development of biofuels options to reduce global CO_2 emissions in the transportation sector, while also meeting a broader set of stakeholder needs in the social and environmental dimensions. Impacts on water resources and biodiversity must be considered, along with regional economic and social factors. Transparency in assessment is critical for effective stakeholder engagement and reliable decision making, to insure the development and deployment of truly value-added energy solutions.

3.1 Overview

The "materiality" of world energy, from supply infrastructure to the quantities of fuel molecules and electrons required to meet demand, is enormous. Most of today's energy supply is derived from hydrocarbons, either fossil sources or renewable bio-based resources. While solar and wind energy solutions have been developed for some applications, cost and convenience continue to limit their broad use in meeting global energy demand. Of an approximate 220 billion tonne per year carbon cycle occurring via natural processes (including human respiration), the addition of 20–30 billion tonnes per year from burning of fossil fuels to meet humankind's growing energy demand is a concern for the potential impacts on climate change. This concern is driving movement to bio-based feedstocks for energy, where the carbon contained

J. B. Powell (✉)
Shell Projects and Technology, Westhollow Technology Center, Houston, TX 77082, USA
e-mail: joe.powell@shell.com

P. M. Ayoub
Shell Alternative Energies, One Shell Plaza, Houston, TX 77002, USA

S. Kapila · G. Y. Buss
Shell Global Solutions, London, UK

A. P. R. Nevill
Shell Alternative Energies, Shell International Petroleum Company Ltd, London, UK

T. Boone et al. (eds.), *Sustainable Supply Chains*,
International Series in Operations Research & Management Science 174,
DOI 10.1007/978-1-4419-6105-1_3,

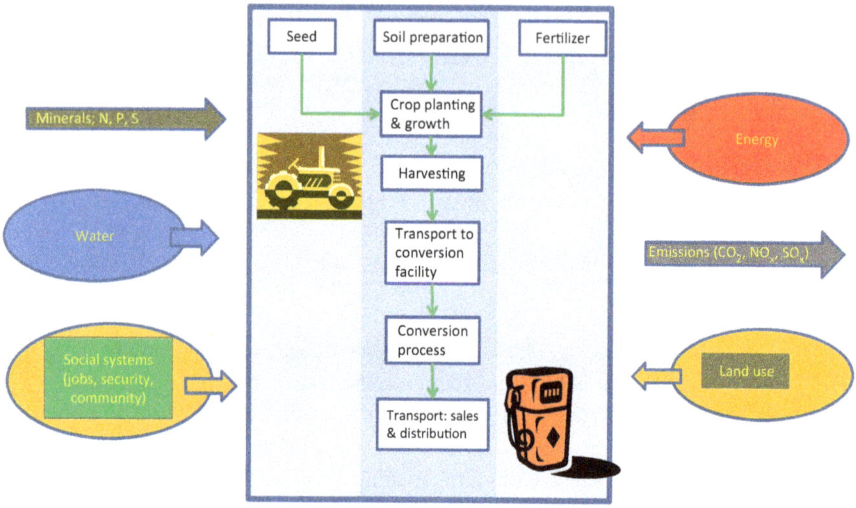

Fig. 3.1 Bio-domain supply chain

in the biofuel is derived via recycle of carbon dioxide (CO_2) from the atmosphere in a closed cycle, driven by energy from sunlight, with storage via photosynthesis.

The enormous scale of world energy demand, measured in terms of the mass of molecules which must be manipulated to generate and transport electricity and fuels, requires astute consideration of the impacts on stakeholders and environment for any energy source and pathway chosen to meet global needs. The current chapter examines the framework required for successful management of supply chains to produce and deliver biofuel s, as a low carbon solution for mitigation of the risk of climate change, targeting primarily the transportation sector. Success must be measured via all three dimensions of sustainability, addressing environmental, social, and economic impacts.

3.2 Biodomain Supply Chain and Stakeholders

Figure 3.1 depicts typical components of the supply chain for production of renewable biofuels, for the case of growth of energy crops vs. use of residues and wastes as the source of bio-based feedstock. Each step within the value chain requires the use of a number of resources. Water, minerals, and land are inputs to primary production. Biomass as a solid feedstock requires truck or rail transport, relative to pipeline for petroleum or natural gas. Biodiversity and impacts on habitats must be considered in conjunction with land use impacts (Tan et al. 2009) associated with biomass production. The social dimension must be considered in terms of impact on local economics, job creation, and quality of work for those so engaged. Non-governmental organizations (NGO's) play a role in helping to develop policies which can support a migration to renewable biofuels. Given the mega-scale infrastructure required for fuels production, clear and sustainable economic incentives must exist over extended periods of time, to provide the necessary stability for investment at levels needed to drive impactful change.

A key metric in deciding between options for feedstock and production routes for biofuels is the net CO_2 impact across the supply chain. "Well-to-wheels" impact on CO_2 reduction, perhaps better stated as field or forest to wheels, must be accurately assessed, and include all sources of energy used in the supply and production chain. This is illustrated in Table 3.1, which was developed using the online tool and database from (US) Argonne National Labs, comprising an introduction to the GREET Model (Greenhouse Gases, Regulated Emissions, and Energy Use in Transportation) and its uses.[1] The source of the energy used in the conversion process thus becomes an important factor in determining whether or not the use of a particular biofuel instead of its fossil energy alternative is indeed helpful in reducing net CO_2 emissions. As the data suggests for this illustration, production of ethanol from lignocellulosic biomass such as corn stover, switchgrass, or forest residues can result in significantly lower greenhouse gas emissions, due in part to the fact that unconverted biomass is also used for the generation of a substantial portion of the energy required by the process, and also for generation of excess electricity that is fed back to the grid, thus displacing the use of fossil fuels in multiple arenas. Improved areal production efficiency for biomass vs. sugars, starches, and oils also contributes to the enhanced ability of biomass to deliver significant volumes of future fuels, with significantly reduced carbon footprint.

3.3 Economics of Sustainability and the Environment

"Sustainability" is defined and measured by the particular stakeholder groups involved in the assessment (Powell 2010). A majority of stakeholders for Western civilization prior to the late twentieth century were primarily concerned with meeting basic needs of food, shelter, and clothing. This remains true for a subset of stakeholders in the "developed" world, and to a greater degree for those nations whose gross domestic product would be characterized as the "developing world". Meeting basic needs continues to be a high priority for much of the world's population.[2]

Feedback mechanisms for basic needs (hunger, cold, inconvenience) are more immediate than those for longer term impacts of environmental stressors (air and water quality). The feedback mechanisms for longer-term CO_2 impacts and climate change at a global level are generally less apparent than local environmental concerns. This creates a fundamental challenge in mechanisms for the global community to deal

[1] ANL GREET Model (http://greet.es.anl.gov/results).

[2] http://en.wikipedia.org/wiki/Developed_country.

Table 3.1 Illustrative example calculation of energy required for the production of one gallon of gasoline equivalent from various sources. (http://greet.es.anl.gov/results)

Fuel		Gasoline			Ethanol				
Feedstock		Conventional crude	Oil sand	US average	Corn	Switchgrass	Corn stover	Forest residue	Sugar cane
Total energy [Btu/gge]	WTP	20,656	39,455	23,256	77,268	122,016	149,646	106,130	141,689
	PTW	113,602	113,602	113,602	113,602	113,602	113,602	113,602	113,602
	WTW	134,257	153,056	136,858	190,869	235,617	263,247	219,731	255,291
Fossil energy [Btu/gge]	WTP	19,834	38,415	22,405	63,888	−5,060	−11,018	−8,941	17,252
	PTW	110,520	110,520	110,520	0	0	0	0	0
	WTW	130,354	148,935	132,925	63,888	−5,060	−11,018	−8,941	17,252
Coal [Btu/gge]	WTP	2,067	2,909	2,184	11,463	−15,750	−16,223	−16,005	1,457
	PTW	0	0	0	0	0	0	0	0
	WTW	2,067	2,909	2,184	11,463	−15,750	−16,223	−16,005	1,457
NG [Btu/gge]	WTP	8,333	26,458	10,841	45,023	3,341	−3,547	−475	5,638
	PTW	0	0	0	0	0	0	0	0
	WTW	8,333	26,458	10,841	45,023	3,341	−3,547	−475	5,638
Petroleum [Btu/gge]	WTP	9,434	9,048	9,381	7,401	7,348	8,752	7,538	10,157
	PTW	110,520	110,520	110,520	0	0	0	0	0
	WTW	119,954	119,568	119,901	7,401	7,348	8,752	7,538	10,157
GHG [gCO2e/gge]	WTP	1,976	3,425	2,177	−596	−7,961	−9,310	−9,292	−5,673
	PTW	8,817	8,817	8,817	8,605	8,605	8,605	8,605	8,605
	WTW	10,793	12,242	10,994	8,009	644	−704	−686	2,932

Results created by ANL on 11/11/2010 using GREET1.8d.1 version, August 2010 release; Energy use results in this tab are provided in units of Btu per equivalent gallons of gasoline (gge) of the consumed fuel; *WTP* well to pump; *PTW* pump to wheels; *WTW* well to wheels with risks of CO_2 and climate change, where stakeholders from different economic backgrounds or perceived quality of life will assign different values and measures to what is the optimal choice for "sustainability".

A key question is "who pays" for the environmental and planetary benefits of reduced carbon energy sources? As summarized in a recent treatise on *Economics and the Environment* (Goodstein 1999), the economic behavior of the majority of consumers as environmental stakeholders is to want environmental burdens to be shared, as opposed to individually paying a premium for "greener" or environmentally advantaged products. This is true even for that segment of global population with wealth beyond what is needed to meet basic needs. The issue is one of fairness in sharing environmental burdens. Thus, cleaner air and water, or a move to renewable fuels to address climate change, are implemented by governmental intervention in the form of environmental permits, as well subsidies or mandates for environmentally advantaged fuels and products.

For the example of sustainable palm oil for biodiesel, Boons and Mendoza describe the connection of the Netherlands with Columbia as the consumer market and supplier in the palm oil supply chain for renewable energy (Boons and Mendoza 2010). Demand for renewable palm oil to address climate change has been driven by governmental policies in the Netherlands and the European Union (EU). Permitting of new palm plantations in Columbia, with net impacts on carbon sinks and sources, is managed by the Columbian government, who must deal with the complexity of assessing indirect land use change (ILUC) impacts on CO_2, as well as other sustainability metrics (Tan et al. 2009). Technology transfer is thus highly important, such that knowledge of impacts of land use change on carbon balances can be transmitted to those engaged in production decision making. Smaller local producers need access to results from available tools for life cycle inventory and assessment, and awareness of how production methods (fertilizer use, harvesting) can impact the carbon intensity, in order to make sustainable decisions with positive benefit on carbon balances (Tan et al. 2009). Validation or certification of carbon intensity of the biofuels product is an important metric for this supply chain.

In order to effect technology transfer and drive optimal environmental and social decisions, The Roundtable on Sustainable Palm Oil (RSPO) was formed in 2004 with the objective of promoting the growth and use of sustainable palm oil products through credible global standards and engagement of stakeholders. The seat of the association is in Zurich, Switzerland, while the secretariat is currently based in Kuala Lumpur. The non-profit association unites palm oil producers, processors or traders, as well as consumer goods manufacturers, retailers, banks and investors, environmental or nature conservation NGOs and social or developmental NGOs—to develop and implement global standards for sustainable palm oil.[3]

In 2010, the EU called for industry, governments and NGOs to set up "voluntary schemes" which certify that biofuels used in the EU meet sustainability criteria. The

[3] http://www.rspo.org/. Accessed 01 January 2011.

"sustainable biofuel certificates" will be policed by independent auditors. EU guidelines for calculating carbon intensity of fuel products have been published.[4] Similarly in the US, the Environmental Protection Agency (EPA) has set up regulations for renewable fuels mandates,[5] which include Renewable Volume Obligations (RVO's) which are managed by tradable Renewable Identification Numbers (RINS) that can be calculated basis the carbon intensity of the supply chain used to manufacture the renewable fuel.[6]

To date, sustainably produced sugarcane ethanol has reduced greenhouse gas emissions more than other competing biofuel supply chains. To ensure continued value-added performance upon further expansion in production, sustainability standards have been established through Bonsucro (formerly the "Better Sugarcane Initiative").[7] In addition, the Roundtable on Sustainable Biofuels (RSB)[8] has been established as an international initiative coordinated by the Energy Center at EPFL in Lausanne that brings together farmers, companies, non-governmental organizations, experts, governments, and inter-governmental agencies to ensure sustainability of the biofuels supply chain. The RSB has developed a third-party certification system for biofuels sustainability standards, encompassing environmental, social and economic principles and criteria through an open, transparent, and multi-stakeholder process.

The above examples highlight mechanisms for ensuring sustainability in biofuels supply chains. Key to success is the creation of mechanisms for technology and information transfer, with transparency in assessment of value and impacts, to ensure that optimal solutions are deployed, considering all three dimensions of sustainability.

3.4 Social Dimension of Sustainability

Much of the published emphasis on "sustainability" relates to climate change and the environment. As already mentioned above, climate change is however not the only dimension of sustainability for biofuels (Plassmann et al. 2010). The social dimension of sustainability often sees less attention than environmental, whereas the third economic dimension of the "triple bottom line" is aggressively managed by market mechanisms, augmented by the aforementioned government policies. The importance of palm oil production for local social and economic development in Columbia is considered alongside the EU's emphasis on reduce carbon intensity fuels, in the analysis of Boons and Mendoza (op cit.). Another example of successful management of both social and environmental aspects of sustainability entails harvesting of sugarcane (Brazil), where mechanized harvesting improves safety, and

[4] http://ec.europa.eu/energy/renewables/biofuels/sustainability_criteria_en.htm.

[5] http://www.epa.gov/otaq/fuels/renewablefuels/regulations.htm.

[6] National Renewable Fuel Standard Program—2010 Compliance. http://www.epa.gov/oms/fuels/renewablefuels/compliancehelp/rfs2-workshop-compliance.pdf. Accessed 01 January 2011.

[7] Better Sugar Cane initiative: http://www.bonsucro.com/.

[8] Roundtable on Sustainable Biofuels (RSB): http://rsb.epfl.ch/.

reduces carbon intensity and hence impact on carbon footprint (Goldemberg et al. 2008).

Supply of basic needs is in general regulated by market forces, and the environmental components via mandates and permitting, or incentives and fines. The social dimension is more difficult to define and regulate. Methodologies such as Total Cost Analysis (AIChE 2000; Lauren 2007), supported in development by the American Institute of Chemical Engineers, attempt to quantify in monetary terms both environmental and social impacts, thus collapsing the sustainability triangle into a single parameter where all impacts are "monetized". Valuations and weighting factors for assigning relative value among various stakeholder groups in general requires direct engagement of those impacted, and are less amenable to publishing in databases.

Plaff and coworkers have addressed the issue of both environmental and social valuations of sustainability in presenting market rules for payments of carbon sequestration via tropical forests (Pfaff et al. 2000). The study proposes a mechanism for valuing carbon sequestration afforded by tropical forests, to bridge the interests of producers and fuel consumers to meet both social and environmental needs.

3.5 Broader Context of Supply Chains and Sustainability

The broader context of sustainable supply chain management beyond the biodomain has been recently reviewed (Seuring and Müller 2008), where the authors again point out the emphasis of "sustainability" publications on "green" and environmental issues, but where treatment of social issues or the integration of all aspects of the sustainability triad, is less common. Two strategies are delineated, the first entailing supplier qualification to avert risks to reputation, where stakeholder engagement is crucial to address concerns. The second entails a more life-cycle-assessment driven policy to deliver sustainable products, where results of the "science" of assessment can vary widely, depending upon the specific background and perspectives of the researchers engaging in the study.

3.6 Summary

A move to replace a substantial portion of fossil hydrocarbon fuels with bio-based renewable feedstocks represents a major shift in resource utilization, and in some cases the underlying social systems required for support. Goals are driven by environmental initiatives to protect biodiversity and reduce carbon intensity as a mitigation for the risk of climate change. Ability to achieve these goals is highly dependent upon the source and pathway used for biofuels production, and must be balanced by consideration of other sustainability issues beyond carbon, which include utilization of water and other mineral resources, impacts upon ecosystems of biobased production, and social impacts for production and transport of a new mega-scale material commodity. The carbon intensity and relative value of the biofuel requires astute assessment of the underlying mechanisms and impacts. A method for crediting and rewarding the most carbon-efficient pathways will be required to achieve the goals of the initiative. Transparency in assessment of footprint, and transfer of that information to all stakeholders involved in decision making and optimization, plays a central role in the development of efficient pathways to a truly renewable energy future.

References

AIChE (2000) Total cost assessment methodology manual: AIChE center for waste reduction technology. Institute for Sustainability (IFS). New York: American Institute of Chemical Engineers. http://www.aiche.org/uploadedFiles/IFS/Products/TCAM.pdf. Accessed April 2009

Boons F, Mendoza A (2010) Constructing sustainable palm oil: How actors define sustainability. J Clean Prod 18(16–17):1686–1695

Goldemberg J, Coelho ST, Guardabassi P (2008) The sustainability of ethanol production from sugarcane. Energy Policy 36(6):2086–2097

Goodstein, ES (1999) Economics and the environment. Prentice Hall, New Jersey

Lauren, L (2007) Sustainability: Keeping the competitive edge. Chem Eng Prog 103(6):44–46

Pfaff ASP, Kerr S, Hughes RF, Liu S, Sanchez-Azofeifa GA, Schimel D, Tosi J, Watson V (2000) The Kyoto protocol and payments for tropical forest: An interdisciplinary method for estimating carbon-offset supply and increasing the feasibility of a carbon market under the CDM. Ecol Econ 35(2):203–221

Plassmann K, Norton A, Attarzadeh N, Jensen MP, Brenton P, Edwards-Jones G (2010) Methodological complexities of product carbon footprinting: A sensitivity analysis of key variables in a developing country context. Environ Sci Policy 13(5):393–404

Powell, JB (2010) Sustainability metrics, indicators, and indices. In: Harmsen J, Powell JB (eds) Sustainable development in the process industries: Cases and Impact. Wiley, Hoboken

Seuring S, Müller M (2008) From a literature review to a conceptual framework for sustainable supply chain management. J Clean Prod 16(15):1699–1710

Tan KT, Lee KT, Mohamed AR, Bhatia S (2009) Palm oil: Addressing issues and towards sustainable development. Renew Sustain Energy Rev 13(2):420–427

Chapter 4
Organizing for Sustainability: Exploratory Analysis of the Healthcare Industry

Tonya Boone

This work was funded in part by the Pioneer Portfolio of the Robert Wood Johnson Foundation.

Abstract Sustainability is fast becoming a business megatrend, essential for organizations to address in order to maintain their competitive positions. Sustainability seeks to balance and simultaneously optimize environmental, social and financial concerns. Sustainability in healthcare represents a particularly challenging undertaking as it necessarily encompasses the wide variety of facilities, operations and activities in a typical healthcare organization. This exploratory study examines the organizational approaches of eight healthcare systems with relatively successful sustainability initiatives.

This study identified several key organizational elements that support the effectiveness of the sustainability initiatives. First, there must be an executive champion who advocates for sustainability to important stakeholders. An executive-level champion is essential in helping to define sustainability for the organizational context, securing resources for sustainability, advocating for sustainability within the organization, linking the sustainability activities to high-level strategy, and working with a sustainability coordinator who oversees day-to-day activities.

Second, the values of the sustainability initiative must be reflected in the organization's high-level mission, vision or values. Linking to the high-level mission helps to underscore the relationship between sustainable operations and patient care. This also defines organizational norms and helps guide decision-making. Finally, performance measurement systems are used to monitor and guide sustainability activities. The organizations studied regularly track a number of key sustainability performance indicators. These are monitored at the highest organizational levels. Performance management systems are also used to align individual, department and facility level activities with strategic goals around sustainability.

T. Boone (✉)
Mason School of Business, College of William and Mary, 23185 Williamsburg, VA, USA
e-mail: Tonya.Boone@mason.wm.edu

T. Boone et al. (eds.), *Sustainable Supply Chains*,
International Series in Operations Research & Management Science 174,
DOI 10.1007/978-1-4419-6105-1_4,

4.1 Introduction

Sustainability is emerging as a competitive imperative for many organizations. Lubin and Esty (2010) identify sustainability as the newest business megatrend which will transform the competitive landscape for many industries[1]. Sustainability balances the economic, environmental and social impacts of an organization's operations. Environmental impacts typically refer to an organization's relation to the natural earth systems, including consumption of non-renewable resources. The social component of sustainability can be more difficult to define. According to the recently crafted ISO 26000 standard on social responsibility[2], social sustainability addresses a business organization's relationship to the communities and societies where it operates. It includes activities around human rights, labor practices, fair operating practices, consumer support, health and safety, and community involvement and development. The challenge for most organizations is how to organize the wide-ranging efforts that support environmental and social sustainability, and effectively integrate them into on-going operations. This study examines how eight healthcare systems organize their sustainability efforts.

4.1.1 Sustainability in Healthcare

While manufacturing firms were among the first to adopt sustainability initiatives, service industries such as healthcare have begun to more aggressively embrace sustainability. Healthcare, as one of the largest service industries, represents a significant opportunity for dramatically affecting economy-wide sustainability performance. The $ 2.5 trillion US healthcare industry represents more than 17% of gross domestic product[3], and its environmental footprint is large and multi-faceted. Healthcare facilities consume enormous amounts of energy. The average hospital uses more total energy than any other commercial building type, and is second only to retail food establishments in energy use per square foot. In 2009, according to the US Department of Energy, the healthcare industry spent over $ 6.5 billion on energy-surpassing every service industry except transportation[4]. In addition, a 2007 study estimated that healthcare accounted for 8% of all US greenhouse gases, and 7% of all

[1] Lubin, D. A. and D. C. Esty, "The Sustainability Imperative," Harvard Business Review, May 2010, vol. 88, no. 5, 42–50.

[2] http://www.iso.org/iso/iso_catalogue/management_and_leadership_standards/social_responsibility/sr_discovering_iso26000.htm. Accessed 20 March 2011.

[3] Truffer et al. "Health Spending Projections Through 2019: The Recession Continues, Health Affairs Journal, March 2010, vol. 29, no. 3, 522–529.

[4] Commercial Buildings Sector Energy Intensities: 1992–2003 and Transportation Energy Consumption Surveys. http://www.eia.doe.gov/emeu/efficiency/cbecstrends/cbecs_tables_list.htm. Accessed 28 December 2010.

carbon dioxide emissions[5]. Healthcare facilities use a multitude of cleaning chemicals, which result in potentially hazardous wastewater effluents. The hazardous chemicals used for anesthetics are often vented, creating potentially hazardous air emissions. Finally, US healthcare facilities collectively generate an estimated 5 million tons of solid waste per year. 5% of this is hazardous.

Hospitals also have significant social impact in their communities. First, hospital employees directly impact their communities. The United States has more than 5,000 hospitals and 595,800 total facilities, employing 14.3 million people[6]. They are the largest employers in many communities. Next, accessibility to healthcare is important for community health. Healthcare provides services that support community wellness, disease prevention and emergency preparedness[7]. Consequently, proximity to and density of healthcare services is a significant indicator of community attractiveness and quality of life. Finally, hospitals offer a range of services that have significant spillover effects on the surrounding communities, including attracting businesses, skilled workers and new residents.

The scope and size of healthcare's social and environmental impact represent tremendous opportunities for the industry. Nevertheless, effectively implementing sustainability in healthcare organizations presents a unique challenge. Healthcare is one of the most complex and rapidly changing industries. Healthcare is continually transformed by new techniques, technologies, pharmaceuticals and delivery systems[8]. Healthcare system managers are under pressure to simultaneously improve cost efficiencies, patient safety and quality. Sustainability in healthcare must address a wide variety of activities while maintaining—or even improving—quality of care. This report examines eight health care systems that have introduced sustainability initiatives and achieved noteworthy results. We examine how these organizations have organized their sustainability initiatives in order to achieve superior performance.

Ray Anderson, a pioneer in business sustainability, characterizes sustainability as a journey, "a long commitment that fundamentally changes an organization." The eight healthcare organizations examined in this study are at different places in their sustainability journeys. Their stories show that there is no single journey that best fits every context. There are insights that we can gain from their experiences. The organizations in this study have developed sustainability initiatives that have reduced their organizations' environmental impact while enhancing the social impact of their activities. They have leveraged sustainability to improve efficiency or reduce costs while maintaining or improving patient quality.

---

[5] Chung, J. W. and Meltzer, D. O., Journal of the American Medical Association, November 2009, vol. 302, no. 18, 1970–1972.

[6] Bureau of Labor Statistics Career Guide to Industries, 2010–2011 Edition www.bls.gov. Accessed 2 January 2011.

[7] http://www.hanys.org/communications/publications/2010/2010-09_hanys_economic_benefit_report.pdf. Accessed 20 March 2011.

[8] Bernstein AB, Hing E, Moss AJ, Allen KF, Siller AB, Tiggle RB. Health care in America: Trends in utilization. Hyattsville, Maryland: National Center for Health Statistics. 2003.

The next section provides an overview of relevant background research on organizational governance. Following that is a description of the study methodology. Next are brief descriptions and case studies of the healthcare organizations participating in the study. The paper concludes with a discussion of the major findings and their implications.

This study uses case study research methods to examine in depth the approaches to leadership, organizational coordination and performance measurement adopted by each system. The organizations are all not-for-profit, and vary in size and geographic location. All are headquartered in the United States.

4.2 Key Components of Organizational Structure

Effective organizational structures align organizational, departmental and individual activities around strategic goals. The key components of organizational structure, examined in this study, include leadership, performance measurement and coordinating mechanisms. This study examines how organizations lead, coordinate and measure the performance of their sustainability efforts. The salient leadership, performance measurement, and coordinating and organizing structures associated with the various sustainability initiatives at each of the organizations in the study are identified.

This study examines three leadership characteristics that potentially affect leadership effectiveness. These include the leader's background and training; leader autonomy; and position formality. Background and training refer to the leader's professional experiences, training and education. Leaders with a high degree of autonomy have a lot of discretion to make decisions about their day-to-day activities, as well as the shape of the sustainability initiative. Low autonomy means that activities are closely monitored and the leader must get approval for most decisions. More autonomy is generally associated with more flexibility and creativity in problem—solving and decision making. Formality specifies the degree to which rules, policies and procedures govern a leader's activities[9]. High formality indicates that the leader complies with proscribed processes in carrying out their work.

Mintzberg (1980), in his synthesis of organizational design research, identified six basic coordinating mechanisms. These mechanisms comprise the ways that organizations integrate the work in the organization. Coordination of sustainability efforts may be carried out using one or more of the following mechanisms: (1) mutual adjustment, which is the informal communication among employees; (2) direct supervision or one person coordinating and monitoring the work of employees throughout the organization; (3) training and education to standardize employee skills; (4) standardization of routines and processes through specified rules and guidelines; (5)

[9] Tatikonda, M. and M. Motoya-Weiss, "Integrating Operations and Marketing Perspectives of Product Innovation: The Influence of Organizational Process Factors and Capabilities on Development Performance," Management Science, 2001, vol. 47, no. 1, 151–172.

standardization of output goals through performance measurement and control systems; and (6) standardization of organizational norms and values via a common mission or vision. Coordinating mechanisms ideally promote intra—organizational collaboration and knowledge sharing. They help to guide and shape employees' interactions.

Performance measurement indicates how organizations measure, analyze and review their performance. It comprises what data is collected, how metrics are identified, and how performance is tracked, reviewed, and used to improve organizational performance. This study examines the performance measurement of the sustainability initiatives. It investigates how sustainability metrics are developed, where they are applied, and how they are used by managers and workers throughout the organizations to promote and improve sustainability performance.

This study does not seek to prescribe the optimal organizational structure choices. Sustainability leadership and coordination will be contingent on the organizational context, for example type of hospitals, history, norms, etc. Hence, the study describes the leadership, performance measurement and coordinating decisions made by these healthcare organizations, along with their organizational contexts. The analysis examines the specific choices in order to identify common factors and establish a framework for decision making and future research.

4.3 Method

This study employs case studies to investigate the organizational structures described above. Case study research is used to investigate the practices of high—performing healthcare organizations. Case study methodology excels when trying to understand complex phenomena in real-world contexts. We focus on organizations that are making significant progress with their sustainability programs and have achieved notable environmental performance within the last 3 years. Eight organizations agreed to be included in the study. A principal informant was identified for each organization and contacted requesting project participation. The principal informants are leaders in the sustainability efforts at their organizations—the sustainability director or coordinator. In-depth interviews with sustainability directors at the hospitals were conducted during November and December 2010.

The interviews were guided by a protocol developed from an extensive review of the research literature. A wide range of literature was reviewed, including work on leadership, coordination, sustainability management and high-performing organizations. The protocol consists of three sections: (1) leadership structure and the background of the primary sustainability leadership, (2) mechanisms for coordinating the sustainability initiative across departments and integrating the activities into overall organization processes, and (3) sustainability performance measurement systems. Additional information was collected through archival data and company documents.

4.4 Findings

The study organizations have adopted a variety of structures for leading, coordinating and measuring the performance of their sustainability initiatives. These approaches are reflective of the differing contexts of the organizations. The following discussion examines the leadership choices, coordinating mechanisms and performance measurement systems more closely, highlighting similarities and differences in the study set.

4.4.1 *Leadership*

The sustainability initiative at each of the organizations relies on a senior-level executive to champion or sponsor sustainability. The executive level champion was often essential in initiating the sustainability initiative or increasing the importance of pre-existing sustainability-related activities. According to one respondent, "the real catalyst was the CEO, who had his own epiphany about relationships between environmental health, human health and our contribution to climate change." According to another participant, a vice president's concern for environmental stewardship was critical for increasing the scale of their initiative. And the sustainability initiative at a third organization, "is really the vision of our CEO ... he has a passion for the environment and connecting the environment to health care. Really, the whole program is his vision and that is driven from the top."

The executive level champion is essential for: (1) securing resources for sustainability activities, (2) advocating for sustainability within and outside of the organization, (3) removing any institutional barriers to sustainability, and (4) overseeing the activities of the sustainability director. The champion and sustainability director work in tandem to lead the sustainability programs.

The sustainability directors are responsible for developing and implementing programs to support high-level sustainability goals and coordinating sustainability activities. The sustainability director is responsible for day-to-day leadership and management of the sustainability initiative. All of the directors have formal position descriptions that specify their activities. Formally defining a sustainability director or manager position was typically a major step in developing the sustainability initiative. In many cases, the directors helped to define the position. At the same time, the directors vary in the degree to which their activities are routinized.

While the directors all have formal goals and performance targets, they were granted relatively high levels of autonomy in achieving their objectives and carrying out their day-to-day activities. This may be due to the dynamic nature of sustainability and the need for organizations to support flexibility and creativity in identifying relevant issues and supporting ongoing improvements.

One of the major issues surrounding a healthcare sustainability director's background is whether the sustainability leader must have a clinical background. The results achieved by these systems show no observed association between the medical background of the leader and the effectiveness of the sustainability initiatives. The leaders come from a variety of backgrounds, including architecture, supply chain and humanities. Most of them, however, do have experience working in healthcare organizations. Those that did not indicated that they felt the need to quickly orient themselves to industry and the organization.

The sustainability director's background, however, did affect the initial focus and the trajectory of the sustainability initiative. Directors with a facilities-related background and training were more likely to focus the initial sustainability activities on the facilities. But the sustainability initiatives did not appear to be constrained by the coordinators' backgrounds, and often benefited from it.

Many of the respondents have training or experience with environmental management systems such as ISO 14001. This seemed to be a critical element, as well as the willingness to engage in ongoing learning with respect to environmental issues. Environmental and social management issues, tools and techniques are constantly emerging, and staying on top of the new knowledge seems to be a critical skill. Most of the managers are involved in a variety of sustainability related business groups, such as Practice Greenhealth and EPA partnership programs.

The participants were also asked what skills they considered essential to successfully leading sustainability in healthcare. To a person, they indicated that communication skills were the most important. First, the sustainability leader must be able to effectively build a strong case for a variety of sustainability initiatives to senior management and external stakeholders. The sustainability coordinator is critical to building that support. Some of the participants were involved in the birth of their respective sustainability initiatives and had to sell the idea to senior management. In addition, sustainability coordinators are often called upon to make a strong case for sustainability investments, particularly when there are no cost benefits or the return-on-investment time frame is relatively long. Sustainability coordinators also need to solicit involvement of employees from all organizational departments. Consequently, they must be able to communicate effectively with highly skilled physicians as well as lower-skilled workers. Finally, coordinators often are the public face of their organization's sustainability program. Thus, they need to be able to communicate with the press and with representatives of local communities.

Several of the participants indicated that process analysis and management skills are an important skill for the leadership role. One participant, for example, brought process analysis skills from the electronics industry to his role as the sustainability director. A number of the participants, especially those with business training, use process tools, such as the plan-do-check-act cycle, or for problem identification and improvement. This highlights the importance of process improvement in managing social and environmental performance.

4.4.2 Coordinating Mechanisms

The initiation and development of sustainability at each of these organizations has been unique. The shape of the initiative is defined by its leadership, coordinating mechanisms, and the organizational context, that is, its history, size, culture, complexity, and norms. This section describes the coordinating mechanisms used and their implications. None of the organizations rely on organization-wide training and education to coordinate their sustainability initiatives. As a result, the discussion focuses on standardization of organization norms through a common vision, direct supervision, mutual adjustment and standardized outputs.

4.4.3 Standardization of Organizational Norms Through a Sustainability Vision

The environmental stewardship and social responsibility values of sustainability are reflected in the missions of the participating organizations. In addition, many of the systems have developed environmental vision statements that articulate an organizational commitment to sustainability. These statements underscore the relationship between hospital operations and patient care. Hospitals, by their physical presence, directly affect the environmental and social health of their surrounding communities. Socially and environmentally healthier communities potentially result in patients who come in with fewer and less severe problems, and they provide positive support to patients discharged to those communities.

Linking sustainability to strategic values has a number of important consequences for coordinating and motivating sustainability-related activities within the organizations. First, strategic values provide a standard for intra-organizational behavioral and cultural norms. High-level values guide employee decision making and activity prioritization. This helps to motivate activities and performance toward common goals.

A mission is also public affirmation about how the organization defines itself. Public promotion of an organization's sustainability must be carefully managed. On the one hand, sustainability may help to bolster the organization's brand and reputation. Research has indicated that intangible assets such as reputation may provide a more enduring competitive advantage than other business assets and support long-term success[10]. Signaling the strategic value of sustainability conveys a powerful message to communities and customers. A reputation for sustainability backed up by demonstrated environmentally and socially responsible behavior and performance may convey a powerful competitive advantage.

On the other hand, healthcare organizations must be careful that their environmentally friendly activities do not undermine perceptions of patient safety and care. In this case, the public's perception of environmentally friendly products and services presents a quandary. A number of studies have found that the public generally perceives such products and services as inferior quality or not as effective as conventional

---

[10] Fombrun, G. J. Reputation: Realizing Value from the Corporate Image. Harvard Business School Press, 1996.

products[11]—although this perception is changing. At the same time, sustainability is often seen as a costly endeavor[12], more like a "luxury" that cost—efficient hospitals dare not indulge in. The reputational effects and public perceptions about sustainability represent a delicate balancing act for healthcare.

A sustainability initiative may help to attract and retain top talent. According to one respondent, their sustainability initiative made them more attractive to potential new hires, especially nurses. A 2009 survey by Kelly Services confirms that workers are more likely to work for socially and environmentally responsible organizations. The survey also indicated that an organization's reputation is very important to workers in all age groups.[13]

The reputational benefits of sustainability are likely to vary across the sample as different geographic regions vary in the level of environmental awareness, which will affect the relative importance of those values to the population. The organizations for the most part do not aggressively promote their sustainability programs and their benefits.

4.4.4 *Direct Supervision*

The organizations all use a sustainability director to oversee sustainability activities and ensure connection to strategic goals. There are many benefits associated with using a central office to coordinate sustainability. The organizations have a wide variety of environmental and social activities spread through many departments and often spread over geographic regions as well. A centralized coordinator serves as a single point of accountability for sustainability efforts. The director can consolidate sustainability knowledge and may serve as an internal consultant for sustainability efforts. The director can identify and deploy internal and external resources to help solve problems. The director brings appropriate people and resources to the tasks and facilitates communication and knowledge sharing. Finally, a director may prevent duplication of effort.

On the other hand, a centralized sustainability director may allow for uneven adoption of a sustainability mindset throughout the organization. Some employees may opt out of sustainability initiatives, and it may be seen as extra work. Indeed, for most of the study organizations, except for a few positions, participation in sustainability initiatives was voluntary.

[11] Wong, V., Turner, W. and Stoneman, P., "Marketing Strategies and Market Prospects for Environmentally-Friendly Consumer Products," British Journal of Management, 1996, vol. 7, no. 3, 263–281.

[12] Natural Marketing Institute 2007 LOHAS Consumer Trends Database, Natural Marketing Institute, 2008.

[13] Kelly Global Workforce Index at www.kellyservices.com accessed January 13, 2011.

4.4.5 Mutual Adjustment

Mutual adjustment refers to coordination via formal and informal communication networks among workers. Much of the communication around sustainability takes place through the green teams that help coordinate and carry out sustainability tasks. Teams allow workers to share knowledge and coordinate sustainability activities directly. They increase worker-to-worker communication and interaction around sustainability.

The teams are used in a variety of ways—generally guided by the sustainability director. Green teams at some of the organizations identify environmental or social problems and develop and implement solutions. Other green teams primarily plan environmental program and activities, such as Earth Day celebrations and recycling events. While these activities are important, some participants acknowledge that these events do not always lead to improved performance or support strategically important goals. Functional and cross-functional teams are used. Functional teams are useful as many of the environmental issues are focused in regional or departmental areas. Cross-functional teams are useful in identifying and addressing system-wide or boundary-crossing environmental issues.

Many of the organizations also use intranets, newsletters, list serves or message boards to facilitate organizational communication. Goals, accomplishments and best practices are shared throughout the organization. The technology is also used as a means of getting new people involved in the green teams and soliciting opportunities and suggestions from workers. Nevertheless, diffusion of sustainability's value and cultural transformation remain a challenge.

4.4.6 Performance Measurement

Performance measurement systems are used to monitor and control the sustainability initiatives. All of the organizations rely on direct measures to assess the effectiveness of their sustainability initiatives. A number of measures are routinely reviewed by senior management; usually, the executive champion monitors performance most closely. The organizations have generally realized cost savings from their sustainability activities, especially those focused on sustainable resource management.

For the most part, these measures focus on resource management and controlling effluents. Energy measures dominate as they are relatively easier to capture and the cost benefits are quickly realized. Many facilities are sub-metered and energy used for different systems, such as HVAC and lighting, are continuously monitored. The organizations also capture a lot of data surrounding waste management systems. Pounds or percentages of regulated, conventional and recycled waste are tracked closely, almost continuously, at most of the facilities. Many of the metrics were developed with the help of Practice Greenhealth and the EPA Energy Star program. Some were already being captured because of regulatory requirements. Increasing recycling and more closely tracking all waste streams were associated with reduced costs. In general, managing resources for sustainability helped the organizations develop critical operational capabilities which could then be applied to other areas.

Sustainability compels organizations to move beyond resource and waste management measures. The social- and community-based benefits often require a longer term to realize benefits. Often these performance benefits may be intangible and difficult to measure. As a result, community and social impacts often go unmeasured. Generally, the respondents felt that performance metrics were an area for ongoing innovation and refinement.

Nonetheless, cost performance pressures dominate the measurement concerns. The US healthcare system is under tremendous pressure to reduce costs. Healthcare costs have increased 300% over the last 20 years, and federal and local governments, as well as consumer advocacy groups, are exerting pressure to contain and control costs. This means that hospitals must understand and communicate how sustainability lowers costs and enhances the quality of care. Sustainable food service, greener cleaners and pharmaceutical programs were most often seen as effectively demonstrating the association between quality, cost and sustainable performance.

4.5 Conclusion

An increasing number of healthcare organizations are implementing sustainability programs. These programs have increasingly become a competitive necessity for and financially beneficial to the institutions that have adopted them as part of their core vision and important organizational value. This study examines eight organizations that have relatively successful sustainability initiatives, in order to describe how they have organized and integrated the variety of sustainability activities. While the study is exploratory, it provides several insights into the prerequisites for advancing sustainability in healthcare organizations. First, an executive champion is essential for the establishment and growth of the sustainability initiative. A champion secures resources and helps to remove institutional obstacles. Next, a sustainability director serves as the hub of sustainability activities—identifying opportunities, bringing workers together, sharing knowledge and enacting tactics to meet sustainability performance targets. The sustainability directors are given a fair amount of discretion in directing and defining the still-evolving sustainability initiatives. The sustainability director and green teams are the primary coordinating mechanisms for these organizations. Other coordinating mechanisms adopted by the organizations include sustainability reflected in the high-level corporate strategy, standardized routines and work processes, and performance measurement systems. Finally, a few of the systems use balance scorecard performance measurement systems to coordinate sustainability activities, align them with strategic goals, and diffuse sustainability responsibilities throughout the organization.

This study also highlights some opportunities to help organizations promote sustainability. Most employee participation in green teams and other activities is voluntary. As a result, participation takes away from required work activities or is undertaken on the employees' own time. Incorporating relevant sustainability activities into position descriptions throughout the organization would further legitimize the work and help to diffuse sustainability throughout the organization. Intra-organizational diffusion is cited as one of the biggest challenges facing these respondents.

There is opportunity to integrate sustainability with other production management approaches, such as Lean and Six Sigma. All are based on a systems view of the organization, and require analytics to continuously improve performance. Healthcare managers have traditionally been reluctant to adopt approaches that originated in manufacturing. They must be carefully adapted for the healthcare environment. However, organizations that have done so have seen tremendous performance, cost and quality improvements. (For example, see *Going Lean in Healthcare* at www.entnet.org.) Managers are starting to recognize opportunities for leveraging the analytical synergies among these systems to simultaneous improve cost, quality and sustainability performance. As an example, some manufacturing organizations simultaneously analyze cost, quality, waste and sustainability in their Value Stream Mapping exercises. And the Baldridge National Quality Program now includes sustainability in its assessment criteria.

Quality and Lean management systems rely on formal, functionally-based teams with trained employees to identify performance improvement opportunities and solve problems. Employees who know their respective processes best are empowered to innovate and improve their areas. Training employees in problem-solving techniques with an environmental and social stewardship focus could accelerate sustainability performance.

The study findings also suggest opportunities for future research. This is a small, exploratory study that sampled only hospitals with well-performing sustainability initiatives. A larger sample could incorporate a broader range of types of hospitals at differing levels of performance. Larger studies could test the associations proposed by the results of this study, such as the associations among leadership choices, coordinating mechanisms and sustainability performance. A larger study could also evaluate the relationship between sustainability and quality performance. As indicated above, quality and sustainability rely on process management. Sustainability and quality performance would be reflective of process management capabilities and competence, and may lead to a positive association between the two.

Chapter 5
Supply Chain Network Design of a Sustainable Blood Banking System

Anna Nagurney and Amir H. Masoumi

Abstract In this paper, we develop a sustainable network design/redesign model for the complex supply chain of human blood, which is a valuable yet highly perishable product. Specifically, we consider the optimal design (or redesign) of a blood banking system consisting of collection sites, blood centers, testing and processing labs, storage facilities, distribution centers as well as demand points. Our multicriteria system-optimization approach on networks with arc multipliers captures several critical concerns associated with blood banking systems including but not limited to the determination of the optimal capacities and the optimal allocations, the induced supply-side risk, and the induced cost of discarding potentially hazardous blood waste, while the uncertain demand for blood is satisfied as closely as possible.

5.1 Introduction

Medical waste, also known as *clinical waste*, refers to the waste products that can not be considered as general waste, and that is produced, typically, at health care premises, including hospitals, clinics, and labs. Due to the potentially hazardous nature of medical waste, both the American Dental Association (ADA) and the Centers for Disease Control (CDC) recommend that medical waste be removed in accordance with regulations (Pasupathi et al. 2011).

It is interesting to note that the health care facilities in the United States are second only to the food industry in producing waste, generating more than 6,600 tons per day, and more than 4 billion pounds annually (Fox News 2011). In addition, according to USA Today (2008), considerable amounts of drugs have been found in 41 million Americans' drinking water due to the improper disposal of unused or expired drugs placed in domestic trash or discarded in the waste water. In other countries, up to 4 pounds of waste per hospital bed per day is produced, out of which 0.5% might be categorized as risky/potentially hazardous waste (The News International 2011).

Disposal of medical waste is not only costly to the health care industry, but also may harm the environment. Consequently, poor management of such waste may lead

A. Nagurney (✉) · A. H. Masoumi
Department of Finance and Operations Management, Isenberg School of Management,
University of Massachusetts, 01003 Amherst, MA, USA
e-mail: nagurney@isenberg.umass.edu

T. Boone et al. (eds.), *Sustainable Supply Chains*,
International Series in Operations Research & Management Science 174,
DOI 10.1007/978-1-4419-6105-1_5,

to the contamination of water, the soil, and the atmosphere. While many hospitals choose to have their waste burned so as to avoid polluting the soil through landfills, the incinerators themselves are one of the nation's leading sources of toxic pollutants such as dioxins and mercury (Giusti 2009; Association of Bay Area Governments 2003). Thus, minimizing the amount of medical waste throughout the health care supply chains will lead to a cleaner environment, which may, in turn, also reduce illnesses and death.

When it comes to blood supply chains, the scarcity and vitalness of this highly perishable health care product make such supply chains crucial. Hence, the effective design and control of such systems can support the health and well-being of populations and can also positively affect the sustainability of the environment by reducing the associated waste. Indeed, since blood waste is a significant hazard to the environment, a major step in attaining a sustainable blood supply chain is to be able to minimize the outdating of blood products while satisfying the demand.

In this paper, we develop a multicriteria *system-optimization* framework for the supply chain network design of a sustainable blood banking system. The framework allows for the simultaneous determination of optimal link capacities through investments, and the flows on various links, which correspond to such application-based supply chain network activities as: blood collection, the shipment of collected blood, its testing and processing, its storage, its shipment to distribution centers, and, finally, to the points of demand. The system-optimization approach is believed to be mandated for critical supplies (Nagurney et al. 2011) in that the demand for such products must be satisfied as closely as possible at minimal total cost. The use of a profit maximization criterion, as in Nagurney (2010a), is not appropriate for an organization such as, for example, the American Red Cross, due to its non-profit status.

In particular, the sustainable supply chain network design model for blood banking developed here is novel for several reasons: (1). it captures the perishability of the product through the use of arc multipliers; (2). it handles the costs associated with the discarding of the medical waste, which could be hazardous, (3). it captures the uncertainty associated with the demand for the product along with the risk associated with procurement of the product, and (4). it allows for total cost minimization and the total risk minimization associated with the design and operation of the blood banking supply chain network.

Our framework is a contribution to the growing literature on sustainable supply chains and to the design of sustainable supply chains, in particular (cf. Nagurney and Nagurney (2010) and the references therein). However, our supply chain network design model for sustainable blood systems focuses not on the minimization of emissions but rather on the minimization of waste. Moreover, it captures the perishability of this product.

Recently, several authors have applied derivations of integer optimization models such as facility location, set covering, allocation, and routing to address the optimization/design of supply chains of blood or other perishable critical products (see Pierskalla 2005; Yang 2006; Sahin et al. 2007; Sivakumar et al. 2008; Cetin and Sarul 2009; Ghandforoush and Sen 2010). Furthermore, inventory management methods (for instance, see Karaesmen et al. 2011), Markov models (Boppana and Chalasani

2007) as well as simulation techniques (Rytila and Spens 2006; Mustafee et al. 2009) have also been utilized to handle blood banking systems. Our multicriteria system-optimization approach is quite general and takes into account such critical issues as the determination of optimal capacities and allocations, the induced supply-side risk, uncertain demand, as well as the induced cost of discarding the waste. Furthermore, our mathematical model can be efficiently solved.

The paper is organized as follows. In Sect. 5.2, we develop the supply chain network design model for a sustainable blood banking system that allows for the design of such a network from scratch or the redesign of an existing network. Interestingly, the total number of Red Cross testing laboratories in the United States has decreased from 7–5 over the past few years, mainly due to the economic situation (Rios 2010). Although the closure of the testing facilities has reduced the overall costs of the American Red Cross, it has increased the transportation costs corresponding to the blood service divisions that are far from the testing labs. Our model enables the reevaluation of such modifications to a blood supply chain network system. We emphasize that Nagurney et al. (2012) proposed an operations management model for blood supply chain networks but did not focus on the more challenging aspect of the design (and redesign) of such supply chains. The notation of our model is based on that model but here we make the crucial extension of including link capacities as decision variables. The formulation and analysis of the model are done through the theory of variational inequalities (see Nagurney 1999), since this enables the creation of a foundation in which other models, including decentralized ones, can then be constructed.

In Sect. 5.3, we propose an algorithmic scheme that yields closed form expressions at each iteration in terms of the product path flows, the link capacities, and the associated Lagrange multipliers. We then apply the algorithm to a spectrum of numerical examples, which illustrate the generality and applicability of our methodological and computational framework. We provide a summary and our conclusions in Sect. 5.4.

5.2 The Sustainable Blood Banking System Supply Chain Network Design Model

In this section, we develop the supply chain network design model for a blood banking system. It is important to mention, at the outset, that our sustainable supply chain network design model is applicable to many perishable products, notably, those associated with health care, as in the case of medicines and vaccines, with minor modifications. For continuity purposes, the notation for our model follows closely that of Nagurney et al. (2012).

For definiteness, please refer to Fig. 5.1. Figure 5.1 depicts a possible network topology of a blood banking system. In this network, the top level (origin) node represents the organization. Every other node in the network denotes a component/facility in the system. A path connecting the origin node to a destination node, corresponding to a demand point, consists of a sequence of directed links which correspond

Fig. 5.1 Initial supply chain network topology for a blood banking system

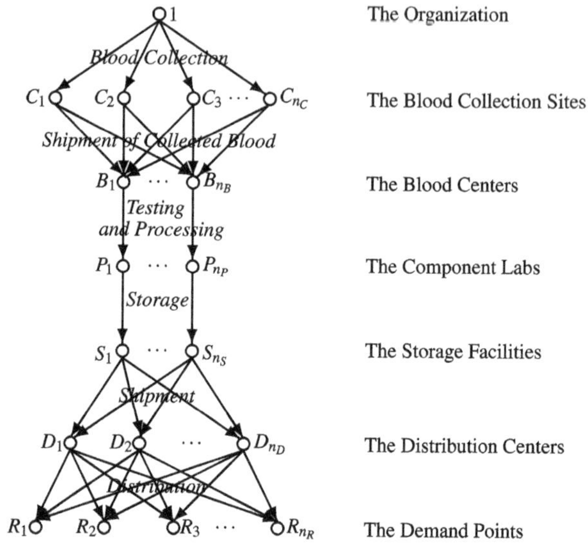

to supply chain network activities that ensure that the blood is collected, processed, and, ultimately, distributed to the demand point. We assume that, in the initial supply chain network topology, as in Fig. 5.1, which serves as a template upon which the optimal supply chain network design is constructed, there exists at least one path joining node 1 with each destination node. This assumption guarantees that the demand at each demand point will be met as closely as possible, given that we will be considering uncertain demand for blood at each demand point. The solution of the model yields the optimal investments associated with the various links as well as the optimal flows, at minimum total cost and risk, as we shall demonstrate and, hence, the optimal sustainable supply chain network design. Our model is sufficiently flexible in that it is capable of handling either the design of the sustainable network from scratch or the redesign of an existing blood banking supply chain network since certain existing link capacities can be either enhanced or reduced.

In the network in Fig. 5.1, we assume that the organization is considering n_C possible blood collection sites constituting the second tier of the network. Many of these collection sites are mobile or temporary locations while others are permanent sites. In the case of drastic shortages; i.e., natural or man-made disasters, the cognizant organizations are likely to need to import blood products from other regions or even other countries, an aspect that is excluded from this model. The first set of links, connecting the origin node to the second tier, corresponds to the process of blood collection where these collection sites are denoted by: $C_1, C_2, \ldots, C_{n_C}$.

The next set of nodes, located in the third tier, consists of the blood centers. There exist, potentially, n_B of these facilities, denoted, respectively, by $B_1, B_2, \ldots, B_{n_B}$, to which the whole blood (WB) is shipped after being collected at the collection sites. Thus, the next set of links connecting tiers two and three of the network topology represents the shipment of the collected blood.

The fourth tier of the network is composed of the processing facilities, commonly referred to as the component labs. The number of these potential facilities is given by n_P. These facilities are denoted by $P_1, \dots, P_{n_P}$, respectively, and are typically located within the blood center locations. As discussed in Nagurney et al. (2012), at these labs, the collected blood is usually separated into parts, i.e., red blood cells and plasma, since most recipients need only a specific component for transfusions. Every unit of donated whole blood—450–500 ml on average—can provide one unit of red blood cells (RBC) and one unit of plasma. In our formulation, what we refer to as the flow of product is, actually, the amount of whole blood (WB) on the first three sets of links. The flow on the links, henceforth, denotes the number of units of red blood cells (RBC) processed at the component labs, which are, ultimately, delivered to the demand points.

The safety of the blood supply is a vital issue for blood service organizations. For example, presently, only five testing labs are operating across the United States, and these labs are shared among 36 blood regions. Only a small sample of every donated blood unit is sent to the testing labs, overnight, and these samples are discarded regardless of the results of the tests. Due to the high perishability of many of the blood products, the two processes of testing and separating take place concurrently yet sometimes hundreds of miles away. If the result of a test for a specific unit of donated blood at the testing lab turns out to be positive, the remainder of that unit will be, subsequently, discarded at the corresponding storage facility. In our model, the set of the links connecting the component labs to the storage facilities corresponds to testing and processing, and the costs on these links represent the operational costs of testing and processing combined. The fraction of the flow lost during or, as a result of the testing process, is also included in our model.

The fifth set of nodes denotes the short-term storage facilities. There are, potentially, n_S of such nodes in the network, denoted by $S_1, S_2, \dots, S_{n_S}$, which are usually located in the same place as the component labs. The links connecting the upper level nodes to the storage facilities denote the procedure of "storage" of the tested and processed blood before it is shipped to be distributed.

The next set of nodes in the network represents the distribution centers, denoted by $D_1, D_2, \dots, D_{n_D}$, where n_D is the total number of potential such facilities. Distribution centers act as transshipment nodes, and are in charge of facilitating the distribution of blood to the ultimate destinations. The links connecting the storage tier to the distribution centers are of shipment link type.

Finally, the last set of links joining the bottom two tiers of the network are distribution links, and they terminate in n_R demand points. The demands at the demand points $R_1, R_2, \dots, R_{n_R}$ are denoted by: $d_{R_1}, d_{R_2}, \dots, d_{R_{n_R}}$, respectively, and the demands are uncertain. Note that, in our design model, the top-tiered node always exists since it represents the organization. Similarly, the bottom-tiered nodes, which correspond to the demand points (such as hospitals and surgical medical centers) also always exist. The solution of our model determines if any of the links should be removed since the optimal solution will yield zero capacities for such links or whether the capacities on links should be increased.

Specific components of the system may physically coincide with some others. Our network model and, hence, the corresponding topology, is process-based rather than location-based, which is compatible with our blood banking problem. Moreover, as mentioned earlier, in general cases of perishable product supply chains, these facilities may be located far apart which can be nicely addressed via our framework.

The possible supply chain network topology, as depicted in Fig. 5.1, is represented by $G = [N, L]$, where N and L denote the sets of nodes and links, respectively. The ultimate solution of the complete model will yield the optimal capacity modifications on the various links of the network as well as the optimal flows.

The formalism that we utilize is that of multicriteria system-optimization, where the organization wishes to determine at what level the blood collection sites should operate; the same for the blood centers, the component labs, the storage facilities, and the distribution centers. Furthermore, the organization seeks to minimize the total supply-side risk as well as the total costs associated with its blood collection, shipment, processing, storage, and distribution activities, along with the total investment corresponding to the enhancement of link capacities (or their construction from scratch), or the total induced cost of reducing link capacities, as well as the total cost of discarding the waste/perished product over the links. The demands must be satisfied as closely as possible with associated shortage penalties if the demands are not met, in addition to the outdating (surplus) penalties in the case that the organization delivers excess supply to the demand points.

With each link of the network, we associate a unit operational cost function that reflects the cost of operating the particular supply chain activity, that is, the collection of blood at blood drive sites, the shipment of collected blood, the testing and processing, the storage, and the distribution. These links are denoted by a, b, etc. The unit operational cost on link a is denoted by c_a and is a function of flow on that link, f_a. The total operational cost on link a is denoted by $\hat{c}_a$, and is constructed as:

$$\hat{c}_a(f_a) = f_a \times c_a(f_a), \quad \forall a \in L. \tag{5.1}$$

The link total cost functions are assumed to be convex and continuously differentiable.

Let w_k denote the pair of origin/destination (O/D) nodes $(1, R_k)$ and let $\mathscr{P}_{w_k}$ denote the set of paths, which represent alternative associated possible supply chain network processes, joining $(1, R_k)$. $\mathscr{P}$ denotes the set of all paths joining node 1 to the destination nodes, and n_p denotes the number of paths.

Let v_k denote the *projected demand* for blood at the demand point $R_k; k = 1, \ldots, n_R$. We assume that the demand at each demand point is uncertain with a known probability distribution. Recall that d_k denotes the actual demand at demand point $R_k; k = 1, \ldots, n_R$, and is a random variable with probability density function given by $\mathscr{F}_k(t)$. Let P_k be the probability distribution function of d_k, that is, $P_k(D_k) = Prob(d_k \leq D_k) = \int_0^{D_k} \mathscr{F}_k(t)d(t)$. Therefore,

$$\Delta_k^- \equiv \max\{0, d_k - v_k\}, \quad k = 1, \ldots, n_R, \tag{5.2}$$

$$\Delta_k^+ \equiv \max\{0, v_k - d_k\}, \quad k = 1, \ldots, n_R, \tag{5.3}$$

where Δ_k^- and Δ_k^+ represent the shortage and surplus of blood at demand point R_k, respectively.

The expected values of the shortage (Δ_k^-) and the surplus (Δ_k^+) are given by:

$$E(\Delta_k^-) = \int_{v_k}^{\infty} (t - v_k)\mathscr{F}_k(t)d(t), \quad k = 1, \dots, n_R, \tag{5.4}$$

$$E(\Delta_k^+) = \int_0^{v_k} (v_k - t)\mathscr{F}_k(t)d(t), \quad k = 1, \dots, n_R. \tag{5.5}$$

As in Nagurney et al. (2012), we associate a relatively large penalty of λ_k^- with the shortage of a unit of blood at demand point R_k, where λ_k^- corresponds to the social cost of a death or a severe injury of a patient, due to a blood shortage. Since blood is highly perishable and will be outdated if not used within a certain period after being delivered, the outdating penalty of λ_k^+ is assigned to the unit of a possible supply surplus. This surplus penalty is charged to the organization because human blood is scarce, and the cognizant organization seeks to minimize the amount of outdated blood at demand points, which actually dominates the amount of blood waste during the other activities of blood banking, for example, within the American Red Cross network (Rios 2010). Hence, in our framework, λ_k^+, in the case of blood (as for other perishable products), includes the cost of short-term inventory holding (cold storage), and the discarding cost of the outdated product. Analogous examples of penalty costs, due to excessive supplies, as well as to shortages, can be found in the literature (see, e.g., Dong et al. 2004; Nagurney et al. 2011). These penalties can be assessed by the authority who is contracting with the organization to deliver the blood.

Thus, the expected total penalty at demand point $k; k = 1, \dots, n_R$, is:

$$E(\lambda_k^-\Delta_k^- + \lambda_k^+\Delta_k^+) = \lambda_k^- E(\Delta_k^-) + \lambda_k^+ E(\Delta_k^+). \tag{5.6}$$

Also, as in Nagurney et al. 2012, we associate with every link a in the network, a multiplier α_a, which corresponds to the percentage of loss over that link. This multiplier lies in the range (0,1], for the network activities, where α_a means that $\alpha_a \times 100\%$ of the initial flow on link a reaches the successor node of that link, with $\alpha_a = 1$, hence, reflecting that there is no waste/loss on link a. For example, the average percentage of loss due to the testing process was reported to be 1.7% (Sullivan et al. 2007); consequently, the corresponding multiplier, α_a, would be equal to $1 - 0.017 = 0.983$.

As mentioned earlier, f_a denotes the (initial) flow on link a. Let f_a' denote the final flow on that link; i.e., the flow that reaches the successor node of the link.

Therefore,

$$f_a' = \alpha_a f_a, \quad \forall a \in L. \tag{5.7}$$

The waste/loss on link a, denoted by w_a, is equal to:

$$w_a = f_a - f_a' = (1 - \alpha_a) f_a, \quad \forall a \in L. \tag{5.8}$$

The organization is also responsible for disposing this waste which is potentially hazardous. Contractors are typically utilized to remove and dispose of the waste. The corresponding discarding cost, y_a, is a function of the waste, w_a, which is charged to the organization:

$$y_a(w_a) = y_a(f_a - f_a') = y_a((1-\alpha_a)f_a), \quad \forall a \in L. \tag{5.9}$$

Since α_a is constant, and known apriori, a new total discarding cost function, $\hat{z}_a$, can be defined accordingly, which is a function of the flow, f_a, and is assumed to be convex and continuously differentiable:

$$\hat{z}_a = \hat{z}_a(f_a), \quad \forall a \in L. \tag{5.10}$$

Let x_p represent the (initial) flow of blood (or a general perishable product) on path p joining the origin node with a destination node. The path flows must be nonnegative, that is,

$$x_p \geq 0, \quad \forall p \in \mathcal{P}, \tag{5.11}$$

since the product will be collected, shipped, etc., in nonnegative quantities.

Let μ_p denote the multiplier corresponding to the loss on path p, which is defined as the product of all link multipliers on links comprising that path, that is,

$$\mu_p \equiv \prod_{a \in p} \alpha_a, \quad \forall p \in \mathcal{P}. \tag{5.12}$$

The projected demand at demand point R_k, v_k, is the sum of all the final flows on paths joining $(1, R_k)$:

$$v_k \equiv \sum_{p \in \mathcal{P}_{w_k}} x_p \mu_p, \quad k = 1, \ldots, n_R. \tag{5.13}$$

Indeed, although the amount of blood that originates on a path p is x_p, the amount (due to perishability) that actually arrives at the destination of this path is $x_p\mu_p$.

As discussed in Nagurney et al. (2012), the multiplier, α_{ap}, is the product of the multipliers of the links on path p that precede link a in that path. This multiplier can be expressed as:

$$\alpha_{ap} \equiv \begin{cases} \delta_{ap} \prod_{a'<a} \alpha_{a'}, & \text{if } \{a' < a\} \neq \varnothing, \\ \delta_{ap}, & \text{if } \{a' < a\} = \varnothing, \end{cases} \tag{5.14}$$

where $\{a' < a\}$ denotes the set of the links preceding link a in path p. Recall that δ_{ap} is defined as equal to 1 if link a is contained in path p; otherwise, it is equal to zero, and $\varnothing$ denotes the null set. In other words, α_{ap} is equal to the product of all link multipliers preceding link a in path p. If link a is not contained in path p, then α_{ap} is set to zero. If a belongs to the first set of links, the blood collection links, this multiplier is equal to 1. The relationship between the link flow, f_a, and the path flows is as follows:

$$f_a = \sum_{p \in \mathscr{P}} x_p \alpha_{ap}, \quad \forall a \in L. \tag{5.15}$$

The blood supply chain organization not only wishes to determine which facilities should operate and at what level, but also is interested in possibly redesigning the existing capacities with the demand being satisfied as closely as possible, and the total cost and risk being minimized. Let $\bar{u}_a$ denote the nonnegative existing capacity on link $a, \forall a \in L$. The organization can enhance/reduce the capacity of link a by $u_a, \forall a \in L$. The total investment cost of adding capacity u_a on link a, or contrarily, the induced cost of lowering the capacity by u_a, is denoted by $\hat{\pi}_a$, and is a function of the change in capacity:

$$\hat{\pi}_a = \hat{\pi}_a(u_a), \quad \forall a \in L. \tag{5.16}$$

The total capacity investment cost functions can be interpreted as the cost of purchasing/renting additional equipments, hiring extra staff personnel, and expanding the transportation fleet. On the other hand, the total cost corresponding to capacity reduction typically includes the relocation of equipment, the reallocation of personnel as well as the storing of surplus equipment. These functions are also assumed to be convex and continuously differentiable. We group the link capacity changes into the vector u. Similarly, the path flows, the link flows, and the projected demands are grouped into the respective vectors x, f, and v.

The total cost minimization objective faced by the organization includes the total cost of operating the various links, the total discarding cost of waste/loss over the links, the total cost of capacity modification, and the expected total blood supply shortage cost as well as the total discarding cost of outdated blood at the demand points. This optimization problem can be expressed as:

$$\text{Minimize} \quad \sum_{a \in L} \hat{c}_a(f_a) + \sum_{a \in L} \hat{z}_a(f_a) + \sum_{a \in L} \hat{\pi}_a(u_a) + \sum_{k=1}^{n_R} \left(\lambda_k^- E(\Delta_k^-) + \lambda_k^+ E(\Delta_k^+)\right) \tag{5.17}$$

subject to: constraints (5.11), (5.13), and (5.15), and

$$f_a \leq \bar{u}_a + u_a, \quad \forall a \in L, \tag{5.18}$$

$$-\bar{u}_a \leq u_a, \quad \forall a \in L. \tag{5.19}$$

Constraint (5.18) guarantees that the flow on a link cannot exceed the new capacity on that link. Furthermore, the change in link capacities can take on positive/negative values corresponding to the enhancement/reduction of the capacities. Constraint (5.19) guarantees that the flow on a link will not be negative by imposing a lower bound for this link capacity change (see Nagurney 2010b).

Observe that if $\bar{u}_a = 0, \forall a \in L$, then the redesign model converts to a "design from scratch" model in that there will be no capacities on the link apriori. Both models of redesign and design are consistent with the presented network topology in Fig. 5.1.

As mentioned earlier, the minimization of total costs is not the only objective of suppliers of perishable goods. A major challenge for a blood service organization is to capture the risk associated with different activities in the blood supply chain network. Unlike the demand, which can be projected according to the historical data, albeit with some uncertainty involved, the amount of donated blood at the collection sites has been observed to be highly stochastic. For example, although blood donors may make appointments in advance, donors may miss their appointments due to traffic delays, bad weather, etc.

As in Nagurney et al. (2012), we introduce a total risk function $\hat{r}_a$ corresponding to link a for every blood collection link. This function is assumed to be convex and continuously differentiable, and a function of the flow, that is, the amount of collected blood, on its corresponding link. The organization attempts to minimize the total risk over all links connecting the first two tiers of the network, denoted by $L_1 \subset L$. The remainder of the links in the network, i.e., the shipment of collected blood, the processing, the storage, shipment, and the distribution links, comprise the set L_1^C. The subset L_1 and its complement L_1^C partition the entire set of links L, that is, $L_1 \cup L_1^C = L$.

Thus, the risk minimization objective function for the organization can be expressed as:

$$\text{Minimize} \sum_{a \in L_1} \hat{r}_a(f_a), \tag{5.20}$$

where $\hat{r}_a = \hat{r}_a(f_a)$ is the total risk function on link a.

The sustainable supply chain network design problem for a blood banking system can be expressed as a multicriteria decision-making problem. The organization seeks to determine the optimal levels of blood processed on each supply chain network link coupled with the optimal levels of capacity escalation/reduction in its blood banking supply chain network activities subject to the minimization of the total cost (operational and discarding) as well as the minimization of the total supply risk. The weight associated with the total cost objective, (5.17), serves as the numeraire, and is set equal to 1. On the other hand, corresponding to the total supply risk objective, (5.20), a weight of θ is assigned by the decision-maker. Thus, the multicriteria optimization problem is:

$$\begin{aligned}\text{Minimize} \quad & \sum_{a \in L} \hat{c}_a(f_a) + \sum_{a \in L} \hat{z}_a(f_a) + \sum_{a \in L} \hat{\pi}_a(u_a) \\ & + \sum_{k=1}^{n_R} (\lambda_k^- E(\Delta_k^-) + \lambda_k^+ E(\Delta_k^+)) + \theta \sum_{a \in L_1} \hat{r}_a(f_a)\end{aligned} \tag{5.21}$$

subject to: constraints (5.11), (5.13), (5.15), (5.18), and (5.19).

The above optimization problem is in terms of link flows. It can also be expressed in terms of path flows:

$$\text{Minimize} \quad \sum_{a \in \mathscr{P}} (\hat{C}_p(x) + \hat{Z}_p(x)) + \sum_{a \in L} \hat{\pi}_a(u_a)$$
$$+ \sum_{k=1}^{n_R} (\lambda_k^- E(\Delta_k^-) + \lambda_k^+ E(\Delta_k^+)) + \theta \sum_{p \in \mathscr{P}} \hat{R}_p(x) \tag{5.22}$$

subject to: constraints (5.11), (5.13), (5.15), (5.18), and (5.19), where the total operational cost function, $\hat{C}_p(x)$, the total discarding cost function, $\hat{Z}_p(x)$, and the total risk function, $\hat{R}_p(x)$, corresponding to path p are, respectively, derived from $C_p(x)$, $Z_p(x)$, and $R_p(x)$ as follows:

$$\begin{aligned}
\hat{C}_p(x) &= x_p \times C_p(x), \quad \forall p \in \mathscr{P}_{w_k}; k = 1, \ldots, n_R, \\
\hat{Z}_p(x) &= x_p \times Z_p(x), \quad \forall p \in \mathscr{P}_{w_k}; k = 1, \ldots, n_R, \\
\hat{R}_p(x) &= x_p \times R_p(x), \quad \forall p \in \mathscr{P}_{w_k}; k = 1, \ldots, n_R,
\end{aligned} \tag{5.23}$$

with the unit cost functions on path p, i.e., $C_p(x), Z_p(x)$, and $R_p(x)$, in turn, defined as below:

$$\begin{aligned}
C_p(x) &\equiv \sum_{a \in L} c_a(f_a)\alpha_{ap}, \quad \forall p \in \mathscr{P}_{w_k}; k = 1, \ldots, n_R, \\
Z_p(x) &\equiv \sum_{a \in L} z_a(f_a)\alpha_{ap}, \quad \forall p \in \mathscr{P}_{w_k}; k = 1, \ldots, n_R, \\
R_p(x) &\equiv \sum_{a \in L_1} r_a(f_a)\alpha_{ap}, \quad \forall p \in \mathscr{P}_{w_k}; k = 1, \ldots, n_R.
\end{aligned} \tag{5.24}$$

Next, we present some preliminaries that help us to express the total shortage as well as the total discarding costs of outdated blood at the demand points solely in terms of path flow variables. Observe that, for each O/D pair w_k:

$$\frac{\partial E(\Delta_k^-)}{\partial x_p} = \mu_p \left[P_k \left(\sum_{p \in P_{w_k}} x_p \mu_p \right) - 1 \right], \quad \forall p \in \mathscr{P}_{w_k}; k = 1, \ldots, n_R. \tag{5.25}$$

Similarly, for the surplus, we have:

$$\frac{\partial E(\Delta_k^+)}{\partial x_p} = \mu_p P_k \left(\sum_{p \in \mathscr{P}_{w_k}} x_p \mu_p \right), \quad \forall p \in \mathscr{P}_{w_k}; k = 1, \ldots, n_R. \tag{5.26}$$

Proof See Nagurney et al. (2012) for the proofs of (5.25) and (5.26). □

We associate the Lagrange multiplier γ_a with constraint (5.18) for link a, and we denote the optimal Lagrange multiplier by γ_a^*, $\forall a \in L$. The Lagrange multipliers may be interpreted as shadow prices. We group these Lagrange multipliers, respectively, into the vectors γ and γ^*.

Let K denote the feasible set such that:

$$K \equiv \{(x,u,\gamma)|x \in R_+^{n_P}, (5.19) \text{ holds, and } \gamma \in R_+^{n_L}\}. \quad (5.27)$$

Before stating the variational inequality formulation of the problem, we recall a lemma that formalizes the construction of the partial derivatives of the total operational cost, the total discarding cost, and the total risk functions with respect to a path flow.

Lemma 1 *The partial derivatives of the total operational cost, the total discarding cost, and the total risk of a path with respect to a path flow are, respectively, given by:*

$$\frac{\partial\left(\sum_{q\in\mathcal{P}}\hat{C}_q(x)\right)}{\partial x_p} \equiv \sum_{a\in L}\frac{\partial\hat{c}_a(f_a)}{\partial f_a}\alpha_{ap}, \quad \forall p\in\mathcal{P}_{w_k}; k=1,\ldots,n_R,$$

$$\frac{\partial\left(\sum_{q\in\mathcal{P}}\hat{Z}_q(x)\right)}{\partial x_p} \equiv \sum_{a\in L}\frac{\partial\hat{z}_a(f_a)}{\partial f_a}\alpha_{ap}, \quad \forall p\in\mathcal{P}_{w_k}; k=1,\ldots,n_R, \quad (5.28)$$

$$\frac{\partial\left(\sum_{q\in\mathcal{P}}\hat{R}_q(x)\right)}{\partial x_p} \equiv \sum_{a\in L_1}\frac{\partial\hat{r}_a(f_a)}{\partial f_a}\alpha_{ap}, \quad \forall p\in\mathcal{P}_{w_k}; k=1,\ldots,n_R.$$

Proof See Nagurney et al. (2012) for the proof. □

We now derive the variational inequality formulation of the problem in terms of path flows and link flows.

Theorem 1 *The optimization problem (5.22), subject to its constraints, is equivalent to the variational inequality problem: determine the vector of optimal path flows, the vector of optimal capacity adjustments, and the vector of optimal Lagrange multipliers $(x^*,u^*,\gamma^*)\in K$, such that:*

$$\sum_{k=1}^{n_R}\sum_{p\in\mathcal{P}_{w_k}}\left[\frac{\partial\left(\sum_{q\in\mathcal{P}}\hat{C}_q(x^*)\right)}{\partial x_p}+\frac{\partial\left(\sum_{q\in\mathcal{P}}\hat{Z}_q(x^*)\right)}{\partial x_p}+\lambda_k^+\mu_p P_k\left(\sum_{p\in\mathcal{P}_{w_k}}x_p^*\mu_p\right)\right.$$
$$\left.-\lambda_k^-\mu_p\left(1-P_k\left(\sum_{p\in\mathcal{P}_{w_k}}x_p^*\mu_p\right)\right)+\sum_{a\in L}\gamma_a^*\delta_{ap}+\theta\frac{\partial\left(\sum_{q\in\mathcal{P}}\hat{R}_q(x^*)\right)}{\partial x_p}\right]$$
$$\times[x_p-x_p^*]+\sum_{a\in L}\left[\frac{\partial\hat{\pi}_a(u_a^*)}{\partial_{u_a}}-\gamma_a^*\right]\times[u_a-u_a^*]$$
$$+\sum_{a\in L}\left[\bar{u}_a+u_a^*-\sum_{p\in\mathcal{P}}x_p^*\alpha_{ap}\right]\times[\gamma_a-\gamma_a^*]\geq 0, \quad \forall(x,u,\gamma)\in K. \quad (5.29)$$

The variational inequality (5.29), in turn, can be rewritten in terms of link flows as: determine the vector of optimal link flows, the vectors of optimal projected demands and the link capacity adjustments, and the vector of optimal Lagrange multipliers $(f^*, v^*, u^*, \gamma^*) \in k^1$*, such that:*

$$\sum_{a\in L}\left[\frac{\partial \hat{c}_a(f_a^*)}{\partial f_a}+\frac{\partial \hat{z}_a(f_a^*)}{\partial f_a}+\gamma_a^*+\theta\frac{\partial \hat{r}_a(f_a^*)}{\partial f_a}\right]\times[f_a-f_a^*]$$
$$+\sum_{a\in L}\left[\frac{\partial \hat{\pi}_a(u_a^*)}{\partial u_a}-\gamma_a^*\right]\times[u_a-u_a^*]+\sum_{k=1}^{n_R}[\lambda_k^+P_k(v_k^*)-\lambda_k^-(1-P_k(v_k^*))]$$
$$\times[v_k-v_k^*]+\sum_{a\in L}[\bar{u}_a+u_a^*-f_a^*]\times[\gamma_a-\gamma_a^*]\geq 0,\quad \forall (f,v,u,\gamma)\in K^1, \tag{5.30}$$

where K^1 *denotes the feasible set as defined below:*

$$K^1\equiv\{(f,v,u,\gamma)\,|\exists x\geq 0, (5.11), (5.13), (5.15), \text{and } (5.19) \text{ hold, and } \gamma\geq 0\}. \tag{5.31}$$

Proof First, we prove the result for path flows (cf. (5.29)). □

The convexity of $\hat{C}_p$, $\hat{Z}_p$, and $\hat{R}_p$ for all paths p holds since $\hat{c}_a$, $\hat{z}_a$, and $\hat{r}_a$ were assumed to be convex for all links a. The convexity of $\hat{\pi}_a$ was also assumed to hold. We just need to verify that $\lambda_k^- E(\Delta_k^-)+\lambda_k^+E(\Delta_k^+)$ is also convex. We have:

$$\frac{\partial^2}{\partial x_{p^2}}\left[\lambda_k^-E(\Delta_k^-)+\lambda_k^+E(\Delta_k^+)\right]=\lambda_k^-\frac{\partial^2E(\Delta_k^-)}{\partial x_{p^2}}+\lambda_k^+\frac{\partial^2E(\Delta_k^+)}{\partial x_{p^2}},$$
$$\forall p\in\mathscr{P}_{w_k}; k=1,\ldots,n_R. \tag{5.32}$$

Substituting the first order derivatives from (5.25) and (5.26) into (5.32) yields:

$$\frac{\partial^2}{\partial x_{p^2}}\left[\lambda_K^-E(\Delta_k^-)+\lambda_k^+E(\Delta_k^+)\right]$$
$$=\lambda_k^-\frac{\partial}{\partial x_p}\mu_p\left[P_k\left(\sum_{p\in\mathscr{P}_{w_k}}x_p\mu_p\right)-1\right]+\lambda_k^+\frac{\partial}{\partial x_p}\mu_pP_k\left(\sum_{p\in\mathscr{P}_{w_k}}x_p\mu_p\right)$$
$$=(\lambda_k^-+\lambda_k^+)(\mu_p)^2\mathscr{F}_k\left(\sum_{p\in\mathscr{P}_{w_k}}x_p\mu_p\right)>0,\quad \forall p\in\mathscr{P}_{w_k}; k=1,\ldots,n_R. \tag{5.33}$$

The above inequality holds provided that $(\lambda_k^-+\lambda_k^+)$, i.e., the sum of shortage and surplus penalties, is assumed to be positive. Hence, $\lambda_k^-E(\Delta_k^-)+\lambda_k^+E(\Delta_k^+)$, and, as a consequence, the multicriteria objective function in (5.22) is also convex.

Since the objective function (5.22) is convex and the feasible set K is closed and convex, the variational inequality (5.29) follows from the standard theory of variational inequalities (see Nagurney 1999).

As for the proof of the variational inequality (5.30), now that (5.29) is established, we can apply the equivalence between partial derivatives of total costs on paths and partial derivatives of total costs on links from Lemma 1. Also, from (5.13) and (5.15), we can rewrite the formulation in terms of link flows and projected demands rather than path flows. Thus, the second part of Theorem 1, that is, the variational inequality in link flows (5.30), also holds. □

Note that variational inequality (5.29) can be put into standard form (see Nagurney 1999) as follows: determine $X^* \in \mathscr{K}$ such that:

$$\langle F(X^*)^T, X - X^* \rangle \geq 0, \quad \forall X \in \mathscr{K}, \tag{5.34}$$

where $\langle \cdot, \cdot \rangle$ denotes the inner product in n-dimensional Euclidean space.

If we define the feasible set $\mathscr{K} \equiv K$, and the column vector $X \equiv (x, u, \gamma)$, and $F(X) \equiv (F_1(X), F_2(X), F_3(X))$, where:

$$F_1(X) = \left[\frac{\partial\left(\sum_{q\in\mathscr{P}} \hat{C}_q(x)\right)}{\partial x_p} + \frac{\partial\left(\sum_{q\in\mathscr{P}} \hat{Z}_q(x)\right)}{\partial x_p} + \lambda_k^+ \mu_p P_k \left(\sum_{p\in\mathscr{P}_{w_k}} x_p \mu_p \right) \right.$$

$$- \lambda_k^- \mu_p \left(1 - P_k \left(\sum_{p\in\mathscr{P}_{w_k}} x_p \mu_p \right) \right) + \sum_{a\in L} \gamma_a \delta_{ap} + \theta \frac{\partial\left(\sum_{q\in\mathscr{P}} \hat{R}_q(x)\right)}{\partial x_p};$$

$$\left. p \in \mathscr{P}_{w_k}; k = 1, \cdots, n_R \right],$$

$$F_2(X) = \left[\frac{\partial \hat{\pi}_a(u_a)}{u_a} - \gamma_a; \quad a \in L \right],$$

and

$$F_3(X) = \left[\bar{u}_a + u_a - \sum_{p\in\mathscr{P}} x_p \alpha_{ap}; \quad a \in L \right], \tag{5.35}$$

then variational inequality (5.29) can be re-expressed in standard form (5.34).

We will utilize variational inequality (5.29) in path flows for our computations since our proposed computational procedure will yield closed form expressions at each iteration. Once we have solved problem (5.29), by using (5.15), which relates the links flows to the path flows, we can obtain the solution f^* which, along with u^*, minimizes the total cost as well as the total supply risk (cf. (5.21)) associated with the design of the supply chain network of a blood banking system.

We now present the algorithm for the solution of the sustainable blood banking supply chain network design followed by several numerical examples.

5.3 The Algorithm and the Numerical Examples

In this Section, we first recall the Euler method, which is induced by the general iterative scheme of Dupuis and Nagurney (1993). Its realization for the solution of the sustainable blood bank supply chain design problem governed by variational inequality (5.29) (see also (5.34)) induces subproblems that can be solved explicitly and in closed form.

Specifically, at an iteration τ of the Euler method (see also Nagurney and Zhang 1996), one computes:

$$X^{\tau+1} = P_{\mathscr{K}}(X^{\tau} - a_{\tau} F(X^{\tau})), \tag{5.36}$$

where $P_{\mathscr{K}}$ is the projection on the feasible set $\mathscr{K}$ and F is the function that enters the variational inequality problem (5.34).

As shown in Dupuis and Nagurney (1993); see also Nagurney and Zhang (1996), for convergence of the general iterative scheme, which induces the Euler method, among other methods, the sequence $\{a_{\tau}\}$ must satisfy: $\sum_{\tau=0}^{\infty} a_{\tau} = \infty$, $a_{\tau} > 0$, $a_{\tau} \to 0$, as $\tau \to \infty$. Specific conditions for convergence of this scheme can be found for a variety of network-based problems, similar to those constructed here, in Nagurney and Zhang (1996) and the references therein. Applications of this Euler method to the solution of oligopolistic supply chain network design problems can be found in Nagurney (2010a).

5.3.1 *Explicit Formulas for the Euler Method Applied to the Sustainable Blood Supply Chain Network Design Variational Inequality (29)*

The elegance of this procedure for the computation of solutions to the sustainable blood supply chain network design problem modeled in Sect. 5.2 can be seen in the following explicit formulas. In particular, (5.36) for the blood supply chain network design problem governed by variational inequality problem (5.29) yields the following closed form expressions for the blood product path flows, the capacity adjustments corresponding to various links, and the Lagrangian multipliers, respectively:

$$x_p^{\tau+1} = \max\left\{0, x_p^{\tau} + a_{\tau}\left(\begin{array}{c} \lambda_k^- \mu_p\Big(1 - P_k\Big(\sum\limits_{p\in\mathscr{P}_{w_k}} x_p^{\tau}\mu_p\Big)\Big) - \lambda_k^+ \mu_p P_k\Big(\sum\limits_{p\in\mathscr{P}_{w_k}} x_p^{\tau}\mu_p\Big) \\ - \dfrac{\partial\Big(\sum_{q\in\mathscr{P}} \hat{C}_q(x^{\tau})\Big)}{\partial x_p} - \dfrac{\partial\Big(\sum_{q\in\mathscr{P}} \hat{Z}_q(x^{\tau})\Big)}{\partial x_p} \\ - \sum\limits_{a\in L} \gamma_a^{\tau}\delta_{ap} - \theta \dfrac{\partial\Big(\sum_{q\in\mathscr{P}} \hat{R}_q(x^{\tau})\Big)}{\partial x_p} \end{array}\right)\right\},$$

$$\forall p \in \mathscr{P}_{w_k}; k = 1, \ldots, n_R; \tag{5.37}$$

Fig. 5.2 The supply chain network topology for the numerical examples

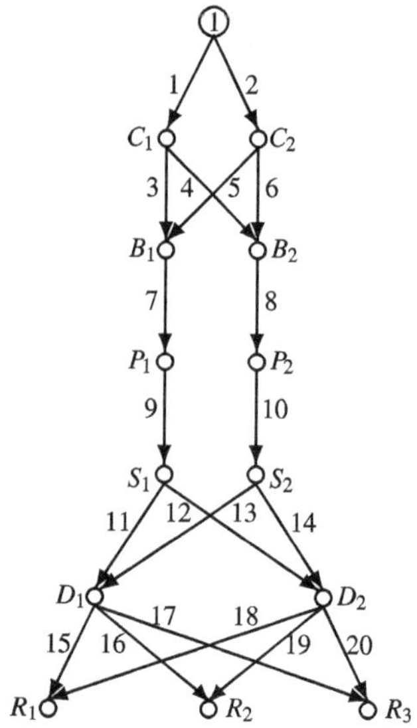

$$u_a^{\tau+1} = \max \left\{ -\bar{u}_a, u_a^\tau + a_\tau \left(\gamma_a^\tau - \frac{\partial \hat{\pi}_a(u_a^\tau)}{\partial_{u_a}} \right) \right\}, \quad \forall a \in L; \tag{5.38}$$

$$\gamma_a^{\tau+1} = \max \left\{ 0, \gamma_a^\tau + a_\tau \left(\sum_{p \in \mathcal{P}} x_p^\tau \alpha_{ap} - \bar{u}_a - u_a^\tau \right) \right\}, \quad \forall a \in L. \tag{5.39}$$

We applied the Euler method to compute solutions to numerical blood supply chain network problems. The initial prospective network topology used in our numerical examples consisted of two blood collection sites, two blood centers, two component labs, two storage facilities, two distribution centers, and three demand points, as depicted in Fig. 5.2. The Euler method (cf. (5.37), (5.38), and (5.39)) for the solution of variational inequality (5.29) was implemented in Matlab. A Microsoft Windows System at the University of Massachusetts Amherst was used for all the computations. We set the sequence $\{a_\tau\} = .1\left(1, \frac{1}{2}, \frac{1}{2}, \cdots\right)$, and the convergence tolerance was $\varepsilon = 10^{-6}$. The algorithm was initialized by setting the projected demand at each demand point and all other variables equal to zero.

Example 1 In this example, we assumed that the existing capacities of all links in the network were zero; hence, the goal was to design a sustainable blood supply chain network from scratch.

We assumed that R_1 was a small surgical center while R_2 and R_3 were large hospitals with higher demand for red blood cells. The demands at these demand points followed the uniform probability distribution on the intervals [5, 10], [40, 50], and [25, 40], respectively. Hence,

$$P_1\left(\sum_{p\in\mathcal{P}_{w_1}} \mu_p x_p\right) = \frac{\sum_{p\in\mathcal{P}_{w_1}} \mu_p x_p - 5}{5}, \quad P_2\left(\sum_{p\in\mathcal{P}_{w_2}} \mu_p x_p\right) = \frac{\sum_{p\in\mathcal{P}_{w_2}} \mu_p x_p - 40}{10},$$

$$P_3\left(\sum_{p\in\mathcal{P}_{w_3}} \mu_p x_p\right) = \frac{\sum_{p\in\mathcal{P}_{w_3}} \mu_p x_p - 25}{15},$$

where $w_1 = (1, R_1), w_2 = (1, R_2)$, and $w_3 = (1, R_3)$.

The shortage and outdating penalties for each of the three demand points—defined by the organization, such as the American Red Cross Regional Division Management—were:

$$\lambda_1^- = 2800, \quad \lambda_1^+ = 50,$$
$$\lambda_2^- = 3000, \quad \lambda_2^+ = 60,$$
$$\lambda_3^- = 3100, \quad \lambda_3^+ = 50.$$

The total risk functions corresponding to the blood collection links were:

$$\hat{r}_1(f_1) = 2f_1^2, \text{ and } \hat{r}_2(f_2) = 1.5f_2^2,$$

and the weight associated with the risk criterion, θ, was 0.7.

The total cost functions corresponding to the capacity adjustment are as reported in Table 5.1. In addition, the multipliers corresponding to the links, the total cost functions, and the total discarding cost functions are also reported there. Most of these numbers have been selected based on the average historical data for the American Red Cross Northeast Division Blood Services (Rios 2010).

Table 5.1 also provides the computed optimal solutions.

As seen in Table 5.1, the optimal capacity on link 18 was zero (and, as expected, so was its flow), which means that D_1 was the only distribution center to serve the demand point R_1. The values of the total investment cost and the cost objective criterion, (5.17), were 42,375.96 and 135,486.43, respectively.

The computed amounts of projected demand for each of the three demand points were:

$$v_1^* = 5.06, \quad v_2^* = 40.48, \text{ and } v_3^* = 25.93.$$

It is interesting to note that for all the demand points, the values of the projected demand were closer to the lower bounds of their uniform probability distributions due to the relatively high cost of setting up a new blood supply chain network from scratch.

Next, we examine the effect of increasing the shortage penalties—while retaining the other costs—with the purpose of reducing the risk of having shortages at our demand points.

Table 5.1 Total cost, total discarding cost, and total investment cost functions, and solution for numerical example 1

Link a	α_a	$\hat{c}_a(f_a)$	$\hat{z}_a(f_a)$	$\hat{\pi}_a(u_a)$	f_a^*	u_a^*	γ_a^*
1	0.97	$6f_1^2+15f_1$	$0.8f_1^2$	$0.8u_1^2+u_1$	47.18	47.18	76.49
2	0.99	$9f_2^2+11f_2$	$0.7f_2^2$	$0.6u_2^2+u_2$	39.78	39.78	48.73
3	1.00	$0.7f_3^2+f_3$	$0.6f_3^2$	$u_3^2+2u_3$	25.93	25.93	53.86
4	0.99	$1.2f_4^2+f_4$	$0.8f_4^2$	$2u_4^2+u_4$	19.38	19.38	78.51
5	1.00	$f_5^2+3f_5$	$0.6f_5^2$	$u_5^2+u_5$	18.25	18.25	37.50
6	1.00	$0.8f_6^2+2f_6$	$0.8f_6^2$	$1.5u_6^2+3u_6$	20.74	20.74	65.22
7	0.92	$2.5f_7^2+2f_7$	$0.5f_7^2$	$7u_7^2+12u_7$	43.92	43.92	626.73
8	0.96	$3f_8^2+5f_8$	$0.8f_8^2$	$6u_8^2+20u_8$	36.73	36.73	460.69
9	0.98	$0.8f_9^2+6f_9$	$0.4f_9^2$	$3u_9^2+2u_9$	38.79	38.79	234.74
10	1.00	$0.5f_{10}^2+3f_{10}$	$0.7f_{10}^2$	$5.4u_{10}^2+2u_{10}$	34.56	34.56	375.18
11	1.00	$0.3f_{11}^2+f_{11}$	$0.3f_{11}^2$	$u_{11}^2+u_{11}$	25.90	25.90	52.80
12	1.00	$0.5f_{12}^2+2f_{12}$	$0.4f_{12}^2$	$1.5u_{12}^2+u_{12}$	12.11	12.11	37.34
13	1.00	$0.4f_{13}^2+2f_{13}$	$0.3f_{13}^2$	$1.8u_{13}^2+1.5u_{13}$	17.62	17.62	64.92
14	1.00	$0.6f_{14}^2+f_{14}$	$0.4f_{14}^2$	$u_{14}^2+2u_{14}$	16.94	16.94	35.88
15	1.00	$0.4f_{15}^2+f_{15}$	$0.7f_{15}^2$	$0.5u_{15}^2+1.1u_{15}$	5.06	5.06	6.16
16	1.00	$0.8f_{16}^2+2f_{16}$	$0.4f_{16}^2$	$0.7u_{16}^2+3u_{16}$	24.54	24.54	37.36
17	0.98	$0.5f_{17}^2+3f_{17}$	$0.5f_{17}^2$	$2u_{17}^2+u_{17}$	13.92	13.92	56.66
18	1.00	$0.7f_{18}^2+f_{18}$	$0.7f_{18}^2$	$u_{18}^2+u_{18}$	0.00	0.00	1.00
19	1.00	$0.6f_{19}^2+4f_{19}$	$0.4f_{19}^2$	$u_{19}^2+2u_{19}$	15.93	15.93	33.86
20	0.98	$1.1f_{20}^2+5f_{20}$	$0.5f_{20}^2$	$0.8u_{20}^2+u_{20}$	12.54	12.54	21.06

Example 2 Example 2 had the exact same data as Example 1 with the exception of the penalties per unit shortage. The new penalties corresponding to the demand points 1, 2, and 3 were as follows:

$$\lambda_1^- = 28000, \quad \lambda_2^- = 30000, \quad \lambda_3^- = 31000.$$

Table 5.2 shows the optimal solution for Example 2; that is, when the shortage penalties are ten times larger than those of Example 1.

A comparison of the optimal capacities in Examples 1 and 2 confirms that raising the shortage penalties, while keeping all operational and investment costs constant, increased the level of activities in all the network links, except for link 18 which stayed inactive. Due to the increased capacities, the new projected demand values were:

$$v_1^* = 9.33, \quad v_2^* = 48.71, \quad \text{and} \quad v_3^* = 38.09.$$

As seen above, unlike Example 1, here the projected demand values were closer to the upper bounds of their uniform probability distributions. As a result, the values of the total investment cost and the cost objective criterion, were 75,814.03 and 177,327.31, respectively, which were significantly higher than Example 1.

Table 5.2 Total cost, total discarding cost, and total investment cost functions, and solution for numerical example 2

Link a	α_a	$\hat{c}_a(f_a)$	$\hat{z}_a(f_a)$	$\hat{\pi}_a(u_a)$	f_a^*	u_a^*	γ_a^*
1	0.97	$6f_1^2+15f_1$	$0.8f_1^2$	$0.8u_1^2+u_1$	63.53	63.53	102.65
2	0.99	$9f_2^2+11f_2$	$0.7f_2^2$	$0.6u_2^2+u_2$	53.53	53.53	65.23
3	1.00	$0.7f_3^2+f_3$	$0.6f_3^2$	$u_3^2+2u_3$	34.93	34.93	71.85
4	0.99	$1.2f_4^2+f_4$	$0.8f_4^2$	$2u_4^2+u_4$	26.08	26.08	105.34
5	1.00	$f_5^2+3f_5$	$0.6f_5^2$	$u_5^2+u_5$	24.50	24.50	50.00
6	1.00	$0.8f_6^2+2f_6$	$0.8f_6^2$	$1.5u_6^2+3u_6$	27.96	27.96	86.89
7	0.92	$2.5f_7^2+2f_7$	$0.5f_7^2$	$7u_7^2+12u_7$	59.08	59.08	839.28
8	0.96	$3f_8^2+5f_8$	$0.8f_8^2$	$6u_8^2+20u_8$	49.48	49.48	613.92
9	0.98	$0.8f_9^2+6f_9$	$0.4f_9^2$	$3u_9^2+2u_9$	52.18	52.18	315.05
10	1.00	$0.5f_{10}^2+3f_{10}$	$0.7f_{10}^2$	$5.4u_{10}^2+2u_{10}$	46.55	46.55	504.85
11	1.00	$0.3f_{11}^2+f_{11}$	$0.3f_{11}^2$	$u_{11}^2+u_{11}$	35.01	35.01	71.03
12	1.00	$0.5f_{12}^2+2f_{12}$	$0.4f_{12}^2$	$1.5u_{12}^2+u_{12}$	16.12	16.12	49.36
13	1.00	$0.4f_{13}^2+2f_{13}$	$0.3f_{13}^2$	$1.8u_{13}^2+1.5u_{13}$	23.93	23.93	87.64
14	1.00	$0.6f_{14}^2+f_{14}$	$0.4f_{14}^2$	$u_{14}^2+2u_{14}$	22.63	22.63	47.25
15	1.00	$0.4f_{15}^2+f_{15}$	$0.7f_{15}^2$	$0.5u_{15}^2+1.1u_{15}$	9.33	9.33	10.43
16	1.00	$0.8f_{16}^2+2f_{16}$	$0.4f_{16}^2$	$0.7u_{16}^2+3u_{16}$	29.73	29.73	44.62
17	0.98	$0.5f_{17}^2+3f_{17}$	$0.5f_{17}^2$	$2u_{17}^2+u_{17}$	19.89	19.89	80.55
18	1.00	$0.7f_{18}^2+f_{18}$	$0.7f_{18}^2$	$u_{18}^2+u_{18}$	0.00	0.00	1.00
19	1.00	$0.6f_{19}^2+4f_{19}$	$0.4f_{19}^2$	$u_{19}^2+2u_{19}$	18.99	18.99	39.97
20	0.98	$1.1f_{20}^2+5f_{20}$	$0.5f_{20}^2$	$0.8u_{20}^2+u_{20}$	18.98	18.98	31.37

Example 3 In this example, we assumed positive capacities for all the activities of the supply chain network. Thus, the problem became one of redesigning an existing blood supply chain network as opposed to designing one from scratch.

The existing capacity for each link, $\bar{u}_a$, was chosen close to the corresponding optimal solution for capacity, u_a^*, in Example 1, as reported in Table 5.3. All other parameters were the same as in Example 1.

As expected, in Example 3, because of the positive link capacities a priori, the computed values of capacity adjustment, u_a^*, were relatively small. Therefore, the optimal Lagrangian multipliers, γ_a^*, which denote the shadow prices of constraints (5.18), $\forall a \in L$, were considerably smaller than their counterparts in Example 1. Furthermore, the respective values of the capacity investment cost and the cost criterion were 856.36 and 85,738.13.

It is also important to note that, for links 14 and 20, the optimal amounts of capacity adjustment were negative, meaning that the existing capacities were slightly higher than the optimal levels given the probability distribution demands.

The Euler method in Example 3 computed the following projected demand values:

$$v_1^* = 6.62, \quad v_2^* = 43.50, \quad \text{and} \quad v_3^* = 30.40.$$

Table 5.3 Total cost, total discarding cost, and total investment cost functions, initial capacities, and solution for numerical example 3

Link a	α_a	$\hat{c}_a(f_a)$	$\hat{z}_a(f_a)$	$\hat{\pi}_a(u_a)$	$\bar{u}_a$	f_a^*	u_a^*	γ_a^*
1	0.97	$6f_1^2+15f_1$	$0.8f_1^2$	$0.8u_1^2+u_1$	48.00	54.14	6.14	10.83
2	0.99	$9f_2^2+11f_2$	$0.7f_2^2$	$0.6u_2^2+u_2$	40.00	43.85	3.85	5.62
3	1.00	$0.7f_3^2+f_3$	$0.6f_3^2$	$u_3^2+2u_3$	26.00	29.64	3.64	9.29
4	0.99	$1.2f_4^2+f_4$	$0.8f_4^2$	$2u_4^2+u_4$	20.00	22.35	2.35	10.39
5	1.00	$f_5^2+3f_5$	$0.6f_5^2$	$u_5^2+u_5$	19.00	20.10	1.10	3.20
6	1.00	$0.8f_6^2+2f_6$	$0.8f_6^2$	$1.5u_6^2+3u_6$	21.00	22.88	1.88	8.63
7	0.92	$2.5f_7^2+2f_7$	$0.5f_7^2$	$7u_7^2+12u_7$	44.00	49.45	5.45	88.41
8	0.96	$3f_8^2+5f_8$	$0.8f_8^2$	$6u_8^2+20u_8$	37.00	41.40	4.40	72.88
9	0.98	$0.8f_9^2+6f_9$	$0.4f_9^2$	$3u_9^2+2u_9$	39.00	43.67	4.67	30.04
10	1.00	$0.5f_{10}^2+3f_{10}$	$0.7f_{10}^2$	$5.4u_{10}^2+2u_{10}$	35.00	38.95	3.95	44.70
11	1.00	$0.3f_{11}^2+f_{11}$	$0.3f_{11}^2$	$u_{11}^2+u_{11}$	26.00	29.23	3.23	7.45
12	1.00	$0.5f_{12}^2+2f_{12}$	$0.4f_{12}^2$	$1.5u_{12}^2+u_{12}$	13.00	13.57	0.57	2.72
13	1.00	$0.4f_{13}^2+2f_{13}$	$0.3f_{13}^2$	$1.8u_{13}^2+1.5u_{13}$	18.00	22.05	4.05	16.07
14	1.00	$0.6f_{14}^2+f_{14}$	$0.4f_{14}^2$	$u_{14}^2+2u_{14}$	17.00	16.90	−0.10	1.81
15	1.00	$0.4f_{15}^2+f_{15}$	$0.7f_{15}^2$	$0.5u_{15}^2+1.1u_{15}$	6.00	6.62	0.62	1.72
16	1.00	$0.8f_{16}^2+2f_{16}$	$0.4f_{16}^2$	$0.7u_{16}^2+3u_{16}$	25.00	25.73	0.73	4.03
17	0.98	$0.5f_{17}^2+3f_{17}$	$0.5f_{17}^2$	$2u_{17}^2+u_{17}$	14.00	18.92	4.92	20.69
18	1.00	$0.7f_{18}^2+f_{18}$	$0.7f_{18}^2$	$u_{18}^2+u_{18}$	0.00	0.00	0.00	1.00
19	1.00	$0.6f_{19}^2+4f_{19}$	$0.4f_{19}^2$	$u_{19}^2+2u_{19}$	16.00	17.77	1.77	5.53
20	0.98	$1.1f_{20}^2+5f_{20}$	$0.5f_{20}^2$	$0.8u_{20}^2+u_{20}$	13.00	12.10	−0.62	0.00

Example 4 Example 4 was another case of redesigning the blood supply chain network, this time with increased demands. The existing capacities, the shortage penalties, and the cost functions were the same as in Example 3.

The new demands at the three hospitals followed a uniform probability distribution on the intervals [10, 17], [50, 70], and [30, 60], respectively. Thus, the cumulative distribution functions corresponding to the above demands were:

$$P_1\left(\sum_{p\in\mathcal{P}_{w_1}}\mu_p x_p\right)=\frac{\sum_{p\in\mathcal{P}_{w_1}}\mu_p x_p-10}{7},\quad P_2\left(\sum_{p\in\mathcal{P}_{w_2}}\mu_p x_p\right)=\frac{\sum_{p\in\mathcal{P}_{w_2}}\mu_p x_p-50}{20},$$

$$P_3\left(\sum_{p\in\mathcal{P}_{w_3}}\mu_p x_p\right)=\frac{\sum_{p\in\mathcal{P}_{w_3}}\mu_p x_p-30}{30}.$$

Table 5.4 reports the corresponding cost functions as well as the computed optimal solution for Example 4.

As seen in Table 5.4, a 50% increase in demand resulted in significant positive capacity changes as well as positive flows on all 20 links in the network, including

Table 5.4 Total cost, total discarding cost, and total investment cost functions, initial capacities, and solution for numerical example 4

Link a	α_a	$\hat{c}_a(f_a)$	$\hat{z}_a(f_a)$	$\hat{\pi}_a(u_a)$	$\bar{u}_a$	f_a^*	u_a^*	γ_a^*
1	0.97	$6f_1^2+15f_1$	$0.8f_1^2$	$0.8u_1^2+u_1$	48.00	65.45	17.45	28.92
2	0.99	$9f_2^2+11f_2$	$0.7f_2^2$	$0.6u_2^2+u_2$	40.00	53.36	13.36	17.03
3	1.00	$0.7f_3^2+f_3$	$0.6f_3^2$	$u_3^2+2u_3$	26.00	35.87	9.87	21.74
4	0.99	$1.2f_4^2+f_4$	$0.8f_4^2$	$2u_4^2+u_4$	20.00	26.98	6.98	28.91
5	1.00	$f_5^2+3f_5$	$0.6f_5^2$	$u_5^2+u_5$	19.00	24.43	5.43	11.86
6	1.00	$0.8f_6^2+2f_6$	$0.8f_6^2$	$1.5u_6^2+3u_6$	21.00	27.87	6.87	23.60
7	0.92	$2.5f_7^2+2f_7$	$0.5f_7^2$	$7u_7^2+12u_7$	44.00	59.94	15.94	234.92
8	0.96	$3f_8^2+5f_8$	$0.8f_8^2$	$6u_8^2+20u_8$	37.00	50.21	13.21	178.39
9	0.98	$0.8f_9^2+6f_9$	$0.4f_9^2$	$3u_9^2+2u_9$	39.00	52.94	13.94	85.77
10	1.00	$0.5f_{10}^2+3f_{10}$	$0.7f_{10}^2$	$5.4u_{10}^2+2u_{10}$	35.00	47.24	12.24	134.64
11	1.00	$0.3f_{11}^2+f_{11}$	$0.3f_{11}^2$	$u_{11}^2+u_{11}$	26.00	35.68	9.68	20.35
12	1.00	$0.5f_{12}^2+2f_{12}$	$0.4f_{12}^2$	$1.5u_{12}^2+u_{12}$	13.00	16.20	3.20	10.61
13	1.00	$0.4f_{13}^2+2f_{13}$	$0.3f_{13}^2$	$1.8u_{13}^2+1.5u_{13}$	18.00	26.54	8.54	32.23
14	1.00	$0.6f_{14}^2+f_{14}$	$0.4f_{14}^2$	$u_{14}^2+2u_{14}$	17.00	20.70	3.70	9.40
15	1.00	$0.4f_{15}^2+f_{15}$	$0.7f_{15}^2$	$0.5u_{15}^2+1.1u_{15}$	6.00	10.30	4.30	5.40
16	1.00	$0.8f_{16}^2+2f_{16}$	$0.4f_{16}^2$	$0.7u_{16}^2+3u_{16}$	25.00	30.96	5.96	11.34
17	0.98	$0.5f_{17}^2+3f_{17}$	$0.5f_{17}^2$	$2u_{17}^2+u_{17}$	14.00	20.95	6.95	28.81
18	1.00	$0.7f_{18}^2+f_{18}$	$0.7f_{18}^2$	$u_{18}^2+u_{18}$	0.00	0.35	0.35	1.69
19	1.00	$0.6f_{19}^2+4f_{19}$	$0.4f_{19}^2$	$u_{19}^2+2u_{19}$	16.00	21.68	5.68	13.36
20	0.98	$1.1f_{20}^2+5f_{20}$	$0.5f_{20}^2$	$0.8u_{20}^2+u_{20}$	13.00	14.14	1.14	2.83

link 18, which was not constructed/used under our initial demand scenarios. The values of the total investment function and the cost criterion were 5,949.18 and 166,445.73, respectively, and the projected demand values were now:

$$v_1^*=10.65, \quad v_2^*=52.64, \quad \text{and} \quad v_3^*=34.39.$$

Example 5 Example 5 was similar to Example 4, but now the demand suffered a decrease from the original demand scenario rather than the increase that we studied in Example 4. The new demand at demand points 1, 2, and 3 followed a uniform probability distribution on the intervals [4, 7], [30, 40], and [15, 30], respectively, with the following functions:

$$P_1\left(\sum_{p\in\mathscr{P}_{w_1}}\mu_p x_p\right)=\frac{\sum_{p\in\mathscr{P}_{w_1}}\mu_p x_p-4}{3}, \quad P_2\left(\sum_{p\in\mathscr{P}_{w_2}}\mu_p x_p\right)=\frac{\sum_{p\in\mathscr{P}_{w_2}}\mu_p x_p-30}{10},$$

$$P_3\left(\sum_{p\in\mathscr{P}_{w_3}}\mu_p x_p\right)=\frac{\sum_{p\in\mathscr{P}_{w_3}}\mu_p x_p-15}{15}.$$

Table 5.5 Total cost, total discarding cost, and total investment cost functions, initial capacities, and solution for numerical example 5

Link a	α_a	$\hat{c}_a(f_a)$	$\hat{z}_a(f_a)$	$\hat{\pi}_a(u_a)$	$\bar{u}_a$	f_a^*	u_a^*	γ_a^*
1	0.97	$6f_1^2+15f_1$	$0.8f_1^2$	$0.8u_1^2+u_1$	48.00	43.02	−0.62	0.00
2	0.99	$9f_2^2+11f_2$	$0.7f_2^2$	$0.6u_2^2+u_2$	40.00	34.54	−0.83	0.00
3	1.00	$0.7f_3^2+f_3$	$0.6f_3^2$	$u_3^2+2u_3$	26.00	23.77	−1.00	0.00
4	0.99	$1.2f_4^2+f_4$	$0.8f_4^2$	$2u_4^2+u_4$	20.00	17.54	−0.25	0.00
5	1.00	$f_5^2+3f_5$	$0.6f_5^2$	$u_5^2+u_5$	19.00	15.45	−0.50	0.00
6	1.00	$0.8f_6^2+2f_6$	$0.8f_6^2$	$1.5u_6^2+3u_6$	21.00	18.40	−1.00	0.00
7	0.92	$2.5f_7^2+2f_7$	$0.5f_7^2$	$7u_7^2+12u_7$	44.00	38.99	−0.86	0.00
8	0.96	$3f_8^2+5f_8$	$0.8f_8^2$	$6u_8^2+20u_8$	37.00	32.91	−1.67	0.00
9	0.98	$0.8f_9^2+6f_9$	$0.4f_9^2$	$3u_9^2+2u_9$	39.00	34.43	−0.33	0.00
10	1.00	$0.5f_{10}^2+3f_{10}$	$0.7f_{10}^2$	$5.4u_{10}^2+2u_{10}$	35.00	30.96	−0.19	0.00
11	1.00	$0.3f_{11}^2+f_{11}$	$0.3f_{11}^2$	$u_{11}^2+u_{11}$	26.00	23.49	−0.50	0.00
12	1.00	$0.5f_{12}^2+2f_{12}$	$0.4f_{12}^2$	$1.5u_{12}^2+u_{12}$	13.00	10.25	−0.33	0.00
13	1.00	$0.4f_{13}^2+2f_{13}$	$0.3f_{13}^2$	$1.8u_{13}^2+1.5u_{13}$	18.00	18.85	0.85	4.57
14	1.00	$0.6f_{14}^2+f_{14}$	$0.4f_{14}^2$	$u_{14}^2+2u_{14}$	17.00	12.11	−1.00	0.00
15	1.00	$0.4f_{15}^2+f_{15}$	$0.7f_{15}^2$	$0.5u_{15}^2+1.1u_{15}$	6.00	5.52	−0.48	0.63
16	1.00	$0.8f_{16}^2+2f_{16}$	$0.4f_{16}^2$	$0.7u_{16}^2+3u_{16}$	25.00	20.68	−2.14	0.00
17	0.98	$0.5f_{17}^2+3f_{17}$	$0.5f_{17}^2$	$2u_{17}^2+u_{17}$	14.00	16.15	2.15	9.59
18	1.00	$0.7f_{18}^2+f_{18}$	$0.7f_{18}^2$	$u_{18}^2+u_{18}$	0.00	0.00	0.00	1.00
19	1.00	$0.6f_{19}^2+4f_{19}$	$0.4f_{19}^2$	$u_{19}^2+2u_{19}$	16.00	14.58	−1.00	0.00
20	0.98	$1.1f_{20}^2+5f_{20}$	$0.5f_{20}^2$	$0.8u_{20}^2+u_{20}$	13.00	7.34	−0.62	0.00

Table 5.5 displays the optimal solution to this example.

As expected, most of the computed capacity changes were negative as a result of the diminished demand for blood at our demand points. Accordingly, the projected demand values were as follows:

$$v_1^* = 5.52, \quad v_2^* = 35.25, \quad \text{and} \quad v_3^* = 23.02.$$

The value of the total cost criterion for this Example was 51,221.32.

5.4 Summary and Conclusions

In this paper, we developed a sustainable supply chain network design model for a highly perishable health care product—that of human blood. The process incorporated the determination of the optimal capacities of the various activities of a blood banking system, consisting of such activities as the procurement of, the testing and processing of, and the distribution of this product. The model has several novel features:

1. it captures the perishability of this life-saving product through the use of arc multipliers;
2. it contains discarding costs associated with waste/disposal;
3. it determines the optimal enhancement/reduction of link capacities as well as the determination of the capacities from scratch;
4. it can capture the cost-related effects of shutting down specific modules of the supply chain due to an economic crisis;
5. it handles uncertainty associated with demand points;
6. it assesses costs associated with shortages/surpluses at the demand points, and
7. it quantifies the supply-side risk associated with procurement.

We illustrated the model through several numerical examples, which vividly demonstrate the flexibility and generality of our sustainable supply chain network design model for blood banking systems. For the sake of generality, and the establishment of the foundations that will enable further extensions and applications, we used a variational inequality approach for both model formulation and solution.

The framework developed here can be applied, with appropriate adaptation, to other perishable products, also in the health care arena, such as medicines and vaccines, as well as to agricultural products, including food.

Acknowledgements This research was supported, in part, by the John F. Smith Memorial Foundation. This support is gratefully acknowledged.

The authors acknowledge Mr. Len Walker, the Director of Business Development for the American Red Cross Blood Services in the greater Boston area, who shared valuable information on the subject. They also thank Dr. Jorge Rios, the Medical Director for the American Red Cross Northeast Division Blood Services, who graciously agreed to be interviewed twice, and who provided enlightening thoughts on the model.

References

Association of Bay Area Governments (2003) Why are hospitals rethinking regulated medical waste management? Environmental Best Practices for Health Care Facilities June 2003

Boppana RV, Chalasani S (2007) Analytical models to determine desirable blood acquisition rates. IEEE international conference on system of systems engineering

Cetin E, Sarul LS (2009) A blood bank location model: A multiobjective approach. Eur J Pure Appl Math 2(1):112–124

Dong J, Zhang D, Nagurney A (2004) A supply chain network equilibrium model with random demands. Eur J Oper Res 156:194–212

Dupuis P, Nagurney A (1993) Dynamical systems and variational inequalities. Annals of Oper Res 44:9–42

Fox News (2011) Going green in the operating room. February 21. Available online at: http://www.foxnews.com/health/2011/02/21/going-green-operating-room

Ghandforoush P, Sen TK (2010) A DSS to manage platelet production supply chain for regional blood centers. Decis Support Syst 50(1):32–42

Giusti L (2009) A review of waste management practices and their impact on human health. Waste Manage 29(8):2227–2239

Karaesmen IZ, Scheller-Wolf A, Deniz B (2011) Managing perishable and aging inventories: Review and future research directions. Planning production and inventories in the extended enterprise, international series in operations research & management acience, vol 151. Springer, New York, pp 393–436

Mustafee N, Katsaliaki K, Brailsford SC (2009) Facilitating the analysis of a UK national blood service supply chain using distributed simulation. Simulation 85(2):113–128

Nagurney A (1999) Network Economics: A variational inequality approach, second and revised edition. Kluwer Academic Publishers, Dordrecht

Nagurney A (2010a) Supply chain network design under profit maximization and oligopolistic competition. Transp Res Part E 46:281–294

Nagurney A (2010b) Optimal supply chain network design and redesign at minimal total cost and with demand satisfaction. Int J Prod Econ 128:200–208

Nagurney A, Masoumi AH, Yu M (2012) Supply chain network operations management of a blood banking system with cost and risk minimization. Comput Manag Sci 9(2):205–231

Nagurney A, Nagurney LS (2010) Sustainable supply chain network design: A multicriteria perspective. Int J Sustain Eng 3:189–197

Nagurney A, Yu M, Qiang Q (2011) Supply chain network design for critical needs with outsourcing. Pap Reg Sci 90(1):123–143

Nagurney A, Zhang D (1996) Projected dynamical systems and variational inequalities with applications. Kluwer Academic Publishers, Boston

Pasupathi P, Sindhu S, Ponnusha BS, Ambika A (2011) Biomedical waste management for health care industry: A review. Int J Biol Med Res 2(1):472–486

Pierskalla WP (2005) Supply chain management of blood banks. In: Brandeau ML, Sanfort F, Pierskalla WP (eds) Operations research and health care, international series in operations research & management science. Kluwer Academic Publishers, Norwell, 70(2):103–145

Rios J (2010) Interviews with the medical director for the American Red Cross Northeast Division Blood Services, Dedham, Massachusetts on July 19, 2010, and December 6, 2010

Rytila JS, Spens KM (2006) Using simulation to increase efficiency in blood supply chains. Manage Res News 29(12):801–819

Sahin G, Sural H, Meral S (2007) Locational analysis for regionalization of Turkish Red Crescent blood services. Comput Oper Res 34:692–704

Sivakumar P, Ganesh K, Parthiban P (2008) Multi-phase composite analytical model for integrated allocation-routing problem—application of blood bank logistics. Int J Logist Econ Glob 1(3/4):251–281

Sullivan M, Cotten R, Read E, Wallace E (2007) Blood collection and transfusion in the United States in 2001. Transfusion 47:385–394

The News International (2011) Experts call for prioritising hospital waste management. March 3. Available online at: http://www.thenews.com.pk/TodaysPrintDetail.aspx?ID=34021&Cat=6&dt=3/3/2011

USA Today (2008) More testing for drugs in water sought. Mendoza, M., March 3. Available online at: http://www.usatoday.com/news/nation/2008-03-16-3533657499_x.htm

Yang X (2006) Choosing transportation alternatives for highly perishable goods: a case study on nuclear medicine. Masters Thesis, Massachusetts Institute of Technology

Chapter 6
From a Caged Tiger to a Sauntering Elephant: Sustainability as a Driver for Profit and Innovation in Indian IT Companies

Vaidyanathan Jayaraman, Mandar Dattatray Malshe and Ajay Anandnarayan

Abstract Sustainability has become a critical market force affecting long-term financial viability and organizational performance. Governments, media, employees and activists are key stakeholders that have become adept at holding companies accountable for the social consequences of their activities. The high-tech information industry has been generally viewed as relatively benign environmentally and has enjoyed a favorable status with respect to the environment. However, as high-tech companies dominate the market place, environmental impacts from IT are attracting more attention. In this book chapter we make a case how sustainability could be viewed as a key driver for profit and innovation in IT companies. We pay special emphasis on India as more and more companies are being forced to confront environmental issues and have to redesign strategies that can recognize the natural limits.

> The Internet and related information technology enabled services hold the promise of rapid, sustainable economic growth that directly benefits everyone on the planet. However, the same forces could also trigger traumatic social, environmental and economic disruption.
>
> Carly Fiorina, Former CEO, Hewlett-Packard

6.1 Introduction

A global UN survey reports that "Never before has world opinion been so united on a single goal as it is on achieving sustainable development." The promulgation of political, economic and social pressures (the so-called triple bottom line) regarding environmental issues over the past several years has caused companies to take a good look into these issues as part of their strategic and operational outlooks. Global sustainability has been defined by the World Commission of Environment and

V. Jayaraman (✉)
Department of Management, School of Business Administration, University of Miami, 33124 Coral Gables, FL, USA
e-mail: vaidy@miami.edu

M. D. Malshe
Business Analyst (Consumer Packaged Goods), Wipro Technologies, Chennai, India

A. Anandnarayan
Business Consultant, Cognizant Business Consulting, Chennai, India

T. Boone et al. (eds.), *Sustainable Supply Chains*,
International Series in Operations Research & Management Science 174,
DOI 10.1007/978-1-4419-6105-1_6,

Development as the ability to "meet the needs of present without compromising the ability of future generations to meet their needs." The inclusion of environmental issues into corporate strategy beyond what is required by government regulation could be viewed as a means to improve a company's match between expectation of stakeholders and the growing environmental concerns (Garrod 1997; Steadman et al. 1995). However, there is no consensus among corporate leaders regarding the specific meaning of enterprise level sustainability (Gladwin et al. 1995); for some, it is a moral mandate while for others it is a legal requirement. Additionally, Hart and Milstein (2003) suggest that "it is also considered as a cost of doing business, a necessary evil to maintain legitimacy and right to operate."

Information Technology (IT) offers solution while creating problems that organizations never envisioned. However, the true impact of IT on growth and productivity continues to be a matter of debate; there is no doubt that the IT sector has been a dynamic one in many developed countries, and India has stood out as a developing country where IT services such as software exports have grown dramatically, despite the country's relatively low level of income and development. An example of IT's broader impact comes from the IT enabled services, a broad category covering many different kinds of data processing and voice interactions that use some IT infrastructure as inputs, but do not necessarily involve the production of IT outputs. However, while there has been an increase in the number and size of IT facilities and people working in these facilities, it has also placed a huge burden on the consumption of natural resources (water, fuel, raw materials) as well as in the quantity of waste that is being generated. In a survey that was conducted by Gartner (Thibodeau 2007), CIOs singled out Green IT as the most important strategic technology that IT companies need to pay attention. Mines (2008) predicted that the Green IT service market will reach nearly $ 5 billion by 2013.

In an emerging economy such as India, the IT industry that is often at the forefront of managerial practice has now become an active player in supporting sustainable economic development. Watson et al. (2010) suggest that "Information systems have been the greatest force for productivity improvement in the last half century." Over the past fifteen odd years, several IT giants including Wipro, TCS, Infosys and HP have progressed through four important stages of environmental sustainability initiatives. This includes:

- ***End-of-Pipeline Approach:*** IT companies took on a reactive posture to environmental management by investing in environmental technologies including end-of-pipeline treatment, end-of-pipeline recovery for use in the same process or elsewhere, efficiency improvements in the production process such as waste minimization and process material substitution and radical redesign of the production process.
- ***Pollution Prevention and Control:*** this implies that organizations continually adapt their production processes to reduce pollution levels to comply with legal requirements—This can be viewed as a cost leadership approach because complying with pollution levels would help companies avoid any penalty or liability cost. Several IT companies made rapid strides in pollution prevention, risk management and making improvements in their operations including a reduction in toxic materials and emissions.

- ***Product Stewardship:*** this is viewed as a form of product differentiation, whereby products and processes are designed in order to minimize any negative environmental burden during the products' life cycle. IT companies established a product stewardship function that focused on developing global supply chain processes to track and manage regulatory compliance issues, customer response systems, public policy matters, product take-back programs, green packaging and design for environment. Additionally, life cycle analysis (LCA) strategies have also been used to quantify the environmental burden that has been created by a product along a supply chain from cradle to grave, selection of materials, production, distribution, consumption and end disposal.
- ***Sustainable Development:*** the goal is to minimize the environmental burden of firm growth through the deployment of clean technologies. Clean technologies includes technologies for pollutant and waste minimization, as well as energy and natural resource efficiency that is applicable to various stages in the production, use and reuse of products and to the provision of service. IT companies viewed the function of clean technology as to eliminate the creation of pollutants from within the production process.

6.2 Laying the Foundation for Sustainability in IT Companies—the Emergence of Green IT

In the early 1990s, IT companies created an Environmental Health and Safety (EHS) group to handle any environmental matters that pertained to their companies—this included interpreting any new environmental policies, auditing divisions for environmental compliance and performance and prompt documentation and reporting to various environmental agencies. The past decade has given birth to the concept of Green IT wherein corporate executives have taken two major views: (a) How can we mitigate the environmental impact caused by the growth in IT—this includes solutions for reducing power consumption of IT equipment, e-cycling and environmentally friendly buildings; (b) How can we use IT to enhance the environment and mitigate the environmental impact of other business processes—this includes solutions such as telecommuting/telepresence and developing geographic information systems for environmental studies.

Hence, the combination of eco-sustainability and the IT infrastructure perspectives can offer important insights in conceptualizing Green IT. For most CIOs and IT vendors, Green IT is about data center efficiencies where organizations consider both technologies and initiatives to reduce power consumption, cooling and real estate costs associated with data center operations (Mines 2008). Another view of Green IT can also include the soft business practices in acquiring, using and disposing IT. Molla et al. (2009) define Green IT as: "*a systematic application of environmental sustainability criteria to the design, production, sourcing, use and disposal of IT technical infrastructure as well as within the human and managerial components of the IT infrastructure in order to reduce IT, business process and supply chain related emissions and waste and improve energy efficiency.*"

Watson et al. (2010) indicates that Green IT addresses issues such as designing energy efficient chips, replacing personal computers with energy efficient thin clients, use of virtualization software to run multiple operating systems on one server, reducing the energy consumption of data centers, reducing electronic waste from obsolete computing equipment and promoting remote computer administration to reduce transportation emissions. As IT organizations began raising data center room temperatures, introducing energy-efficient designs, and streamlining their server environments through consolidation and virtualization, they realized that what had originally been thought of as a risk reduction activity could also result in significant cost savings because data center energy consumption can account for up to 10% of a company's energy bill.

6.2.1 Saving Money with Green IT

In the last decade or so, IT companies have realized that cost savings due to deploying Green IT strategies weren't isolated just to the data centers. The personal computer (PC) environment also wasted energy from computers left on at night and over weekends. At the time, the paradigm was to leave computers running at night to perform maintenance during off-peak hours. This paradigm shifted with the introduction of new software that automatically shut down desktop computers and other IT peripheral devices at night and on weekends while still allowing for scheduled maintenance, saving companies on average between $ 88 and $ 144 per PC. Multiplied across a company's entire PC environment, this savings proved to be substantial. Printing was identified as a third area of IT cost savings. In the United States, office workers use, on average, 10,000 sheets of copy paper each year. In addition, many offices contain a mix of personal and network printers, generally with little thought to how much energy is consumed across the total printing environment. Costs associated with paper, ink, and energy consumption became another opportunity for savings.

IT organizations began to perform assessments of the printing environment, replacing personal printers, faxes, and copy machines with fewer and more energy-efficient multifunction printing devices. They also set default print settings to black and white to reduce the demand for expensive color ink, and they changed to double-sided printing to cut paper use in half. These changes saved companies between 20 and 60%.

6.2.2 Using Green IT for Competitive Advantage

While companies were looking for risk reduction and cost savings, technology manufacturers were quickly achieving competitive advantage through their green practices. Hewlett-Packard and Dell both began producing ultra–energy efficient computers that contained fewer toxic materials. They also simplified packaging, reducing waste and shipping costs at the same time. Software manufacturers also began to understand the market opportunity and produced energy management software that allowed companies to capture a comprehensive energy profile for the entire IT environment by monitoring all of the devices on the network. The result of the progress made in the first few years after Gartner's study was that by 2010, the IT industry as a whole realized **more than $ 2 billion per year in energy savings**. At the same time, an entirely new industry emerged around the development of energy-efficient solutions that allow companies to increasingly fine-tune and optimize their IT environments. Over the past five years, IT has aggressively reduced energy consumption and become more sustainable in the process. We now present some key drivers for Green IT initiatives that organizations must pursue. They include:

- ***Corporate Social Responsibility:*** We define CSR as actions that appear to further some social good that is beyond the interests of the firm and that which is required by law. In IT companies, targets are now included for reducing the impact of their policies on the environment—this includes equipment, people and infrastructure that constitutes a significant footprint—hence identifying and reducing its impact is an important objective.
- ***Cost Reduction Initiatives:*** In the IT industry, because significant spending occurs by way of energy and fuel costs together with cost and operation of IT infrastructure, companies need to pursue strategies to reduce power consumption by IT equipment. Additionally, energy efficient lighting, alternative energy sources and telecommuting can all help in reducing the overall cost of operations.
- ***Creating New Market Opportunities:*** Changing environmental conditions usually creates new market and business opportunities. For IT companies, such opportunities can include promoting environmentally friendly products such as low power hardware, offering services for setting and attaining environmental impact standards and streamlining operational processes for increased effectiveness. One recent initiative is to look at the performance per watt of a server as opposed to raw performance.
- ***Regulatory Standards:*** Environmental policies can be broadly classified as regulatory practices (permits), economic (benefits and incentives for adherence) and informational (audits, product labeling, environmental reporting). Green IT awareness is now forcing senators to promote several legislations along all these issues. Producing energy star appliances and enforcing optimal levels of carbon footprint of IT have become important requirements for obtaining environmental certification.
- ***Creating Awareness:*** Another key driver to Green IT adoption is to create awareness among key stakeholders including clients, employees and senior administrators the urgency to deal with environmental issues. IT companies now employ screen saver messages to educate their employees on the simple things that they can practice (turn monitors off at the end of the work day) that might have a huge impact on the environment. IT companies are also using a desktop power meter that would make individuals aware of their power consumption and use an online eco-footprint calculator to enable associates to gauge their eco-friendliness and identify any scope for improvement. The challenge now is to build a team of "internal" champions who can spread the word and educate people within the company.

We now discuss some of the sustainability initiative that IT companies have pursued through the LEED initiative.

6.2.3 What is a Green Data Center?

A data center (DC) is a facility that houses a large number of electronic equipment (computers, storage devices etc.). Data centers are used by organizations for the purpose of handling the data necessary for their IT operations. IT operations are crucial for the business continuity of the organizations; companies rely on their information systems to run their operations. It is essential to provide a reliable infrastructure for IT operations to store, handle and process organizational data because if these systems become unavailable (known as down time), company operations may be impaired or stopped completely. Green Data center is defined as a DC that uses the optimal quantity of energy to deliver maximum IT productivity. Green DCs are a combination of technology, process and people in addition to using energy saving tools and an optimal level of cooling. Green DCs must create a change management methodology to transform DC technology, process and people to a low carbon future.

As organizations become more global, it has become essential for companies to run their data centers round the clock to ensure availability of essential data as and when required. This is achieved through significant computational redundancy of the hardware used, leading to substantial wastage of power and an increase in the organization's carbon footprint.

The first decade of the twentyfirst century saw rapid growth and change for data centers globally. In India, data center capacity is growing at a CAGR of 31% and stood at 4.25 million square feet at the end of 2010[1]. In this period, data center managers were forced to react to rapid, continuous changes dictated by the capacity and availability requirements of their organizations, and the density of the equipment being deployed to meet those requirements. As the requirement for data storage increased, data centers became more compact and more powerful, thus generating more heat per square foot of data center space.

A data center's environment has to be controlled in order to ensure optimal performance; the recommended temperature range is 16–24°C (61–75°F) and humidity range of 40–55% with a maximum dew point of 15°C. Due to the tropical climate in India, companies rely on air conditioning to keep the data centers at the right temperature. Hence, data centers in India consume more power on an average, because more amount of power is required to operate, cool and maintain them vis-à-vis colder countries. It is estimated that data centers and their servers consume about 23% of power in an IT organization[2] and that energy costs which are now at 10% of an average IT budget could escalate to 50% in the future if no significant steps are taken

[1] http://www.gartner.com/it/page.jsp?id=808212

[2] Steve Kleynhans, VP Computing, Gartner presentation "the Green PC Environment"—presentation, New York, November 2007

to arrest the trend[3]. Traditional data centers' efficiency statistics show dismally low figures: an average DC utilizes only 6% of its servers and only 50% of its facilities[4] and most DCs run at 15% of capacity or less, idling the rest of the time but still consuming electricity. This, along with the associated environmental cost of running traditional data centers has paved the way for the adoption of green data centers in India.

A green DC is constructed in such a way as to store, manage and disseminate data with maximum energy efficiency and minimal environmental impact. Green DCs not only save energy, but also reduce the need for expensive infrastructure upgrades to deal with increased power and cooling demands. A key aspect of DC greening is consolidation. As a part of DC consolidation, organizations remove under-utilized data centers and distribute their work load over the remaining data centers. This not only increases the efficiency of the remaining data centers, but also saves energy and consequently reduces the carbon foot print of the organization due to reduced operational and cooling costs. It has been estimated that consolidation can raise the utilization of a large DC from 5% to as much as 20% and adoption of this technology in India is forecasted to 29% by end of 2011. Consolidation and DC virtualization by TCS, one of the IT giants of India resulted in 79% savings in power consumed by their servers[5].

Power management systems are one of the important constituents of a green data center. It has been estimated that only about one-quarter of the energy coming into a data centre actually goes to the servers and that taking full advantage of power management features and systems can cut data center energy requirements by as much as 20% and increase a data center's Power Usage Efficiency (PUE)[6]. This is also known as the Green Grid standard. Determining and improving this metric will help an IT organization create a efficient computing model. PUE is the ratio of total facility power over total IT equipment. Here total facility power is the total power being provided to the DC. Total IT equipment power is the power used by the IT equipment in the DC and includes storage equipment, servers, network equipment, laptops, desktops, switches and display units.

Consequently, several Indian companies with large data centers are seeking expert help in the domain of power management. For example, Emerson, a leader in the field of precision cooling and power management will help Datacraft, a leading IT solutions and services provider in India build six data centers with optimized energy efficiency through advanced power management solutions[7].

---

[3] Ganesh Mahabala, Regional Director, India and SAARC, VMware "The How and Why of Green DataCenters"

[4] Data Quest India, May 2008. http://www.mckinsey.com/clientservice/ccsi/pdf/Data_Centers.pdf

[5] www.ciio.in/case-study/server-utilization-drives-power-savings-tcs

[6] IDC: Virtualization and Multicore Innovations Disrupt the Worldwide Server Market"—March 2007

[7] http://www.emerson.com/en-US/newsroom/news-releases/emerson-business-news/Pages/Emerson-to-Design-Six-Data-Centers.aspx

6.2.3.1 How to Make the DC Greener?

By consolidating multiple applications on the "least" number of servers provides numerous benefits that include increased server utilization, redundancy elimination, increased performance and decreased maintenance cost. Narasimhan and Raghuraman (2010) provide a five step process to move towards a low carbon DC. This includes:

- ***Take Stock of the Current IT Inventory in the Data Center***—For each equipment that an IT company is operating, it is important to identify and document the supplier, the technology and its associated applications that are running on this equipment. A manager of the data center should have a clear idea on how the legacy environment has been built for the DC. Issues such as date of manufacture, how many hours has a server logged, the energy productivity of the DC are all important pieces of information to track for the manager. Applications that have not been used effectively in a server and are redundant need to be eliminated. Wipro for example offers a comprehensive Application Rationalization Solution (ARS) that identifies opportunities to eliminate, migrate and modernize the application inventory while helping to build a repository of application inventory
- ***Quantify the Key Performance Indicators for the Green Data Center***—Chillers and computer room air conditioners (CRAC) all consume a great deal of energy moving air around to maintain a cool temperature around the DC. Hence it is critical to quantify some of the KPIs for the green DC that includes the overall performance of IT and the energy performance of the DC. Power usage effectiveness, work load analysis of the IT nodes, Data Center Energy productivity and Data center efficiency are all key KPIs that a green data center should track to remain sustainable.
- ***Strategize between Short-term and Long term Horizon***—Fixing the time horizon is critical for any DC. Since a typical DC deals with end-to-end server, storage and network technology services that can undergo constant churn and change, the DC must devise a comprehensive set of metrics that are differentiated between short term and long term issues. For example, for a typical DC, energy conservation can be achieved by using occupancy sensors, auto vents and building management sensors. This takes a reasonable amount of time to achieve and hence should be considered as a long term strategy.
- ***Identify Green IT "Opportunities"***—A manager should seek opportunities that can increase the efficiency of a DC. Usage of an efficient demand management and capacity planning process so that the DC will possess the right processing power and storage is an important decision to pursue. Additionally, by selecting the right suite of solutions to manage a DC within the prescribed boundaries, this will ensure that energy costs savings are maximized on one hand while any greenhouse gas emissions are minimized on the other hand.
- ***Implement Best Practices***—Virtualization is a technique that can increase server utilization while making the DC leaner. By running virtualization software application, the manager can run multiple applications on just one server. Wipro, for

example, recommends "Thin Provisioning", a green DC storage that substantially decreases disk capacity, energy and floor space costs while drastically alleviating storage and system administration on any storage area network. By deployment of monitoring tools, the process of measuring the related metrics that can then lead a DC to its key performance indicators is another best practice that needs to be implemented.

We now move on to discuss how IT companies can practice sustainability through LEED programs.

6.3 Sustainability Through LEED

Leadership in Energy and Environmental Design (LEED) is an internationally recognized certificate given by the US Green Building Council (USGBC), which verifies that a building or community was built with strategies aimed at sustainability involving energy efficiency, reduced CO_2 emissions and environmental quality. In India, the rating is provided by Indian Green Building Council (IGBC). LEED India was started in 2007 and has now been upgraded to 'LEED 2011 for India' to better suit the Indian scenario. LEED India has two programs LEED® India New Construction for buildings in control of the owner or developer and LEED® India Core and Shell buildings for rented or leased spaces. LEED ratings are awarded based on a 100 point scale and can be ranked as certified, silver, gold and platinum with certified being the lowest and platinum being the highest.

Many of the LEED certified buildings in India are owned by IT companies. Well known IT companies which own LEED rated buildings in India include Microsoft, Cisco, Oracle, Wipro, TCS, Infosys, Patni, Cognizant, HCL, and Netmagic. IT services major—WIPRO is the frontrunner in this initiative. Out of the 25 green buildings that belong to the IT industry, 14 belong to WIPRO. Recently Patni and Infosys have constructed LEED certified services. Patni Knowledge Centre in Noida, a green IT center is the largest LEED Platinum rated building outside USA and the second largest in the world. 75% of the area receives natural daylight. 100% of sewage is recycled and air quality is monitored by way of CO_2 sensors.[8] Infosys BPO Limited has been awarded the "Platinum" rating for its building in Jaipur, India. Microsoft India's development center at Hyderabad has a gold rated LEED campus.[9] Wind energy major—Suzlon's headquarters at Hadapsar called Suzlon One Earth was awarded the LEED platinum rating in 2010. The facility uses 30% of the 150 KW power that it generates to feed its IT servers.[10]

Clients of a global industry such as IT services expect energy efficient working conditions that meet global standards. The IT industry has been viewed by many

---

[8] http://www.patni.com/media/30338/Patni-LEED-Certification.pdf

[9] http://www.microsoft.com/india/msidc/life/campus.aspx

[10] http://www.indianexpress.com/news/Suzlon-second-after-Google-to-use-clean-energy-to-power-its-IT-server/617666

as consuming a lot of energy and such "green" investments are seen as attempts to reduce operational costs and energy consumption. Five of India's largest IT service providers are driving sustainability into IT by two ways; involving sustainability in every stage of IT services delivery and by showing clients their efforts in sustainability and educating them (Tarafdar and Modi 2010). An important component in this drive to sustainability is having energy—efficient buildings. Investments in LEED rated buildings in India are primarily done by the major IT players than smaller IT players because a LEED rated building costs anywhere between 15–30% higher than a normal building and additional costs include those paid to consultants thanks to the heavy paper work involved. The return on investment for Green buildings is between 3–5 years based on the type of rating and such buildings consume 20–30% less water and 50% less energy.[11]

The IT services industry thrives on data centers and there is increasing pressure on the data center industry to provide LEED compliant data centers. Also a LEED data center is seen as having a competitive edge over others. IT infrastructure provider Netmagic's facility in Chennai is India's first gold rated LEED data center. While the investment involved was a little less than 10% higher than required for a conventional data center, the company is already seeing a savings of 20% in energy and 50% in water.[12] HCL Infosystems was awarded the LEED platinum rating for its data center in Noida in February 2011. It became the first company to be awarded this rating for internal construction of the data center.[13] With the increased awareness of this rating, a slew of associated companies involving in consultancy and certification are coming up. More IT companies will want to embrace LEED given its advantages of saving energy and costs and competitive positioning in the market. Also, the green design field is changing rapidly and new technologies, products and designs will emerge forcing changes in the rating system.[14]

6.4 E-waste Management

In this section of the chapter, we move onto another major environmental initiative that has plagued the IT/Electronic industry—E-waste management. The electronic industry is the world's largest and fastest growing manufacturing industry. In the last fifteen odd years or so, the industry has provided a big leverage to the socio—economic and technological growth of a developing society. But as every big change comes with effects, the consequence of its consumer oriented growth combined with rapid product obsolescence and technological advances are a new environmental challenge—the growing menace of "Electronics Waste" or "e waste" that consists of obsolete devices dump. The problem can be seen both from the perspective of an

[11] http://dqindia.ciol.com/content/GreenIT/2010/110090401.asp

[12] http://www.netmagicsolutions.com/new-datacenter-launch.html

[13] http://www.hclinfosystems.in/news206.htm

[14] http://www.wiprolighting.com/uploads/greenforce/LEED-India%20CS%20guide.pdf

issue and from the perspective of a business opportunity of increasing significance, given the volumes of e-waste being generated and the content of both toxic and valuable materials in them. The fraction including iron, copper, aluminum, gold and other metals in e-waste is over 60%, while plastics account for about 30% and the hazardous pollutants comprise only about 2.70%.

India in addition to dealing with solid waste management, has to also deal with environmental problems surrounding E-waste. India has been a popular destination of e-waste dumping and e-waste from developed countries find an easy way into developing countries in the name of free trade and is further complicating the problems associated with waste management. In this section of the chapter, we highlight the associated issues and strategies to address this emerging problem, in the light of initiatives in India.

6.4.1 Electronic Waste

Electronic waste or E-waste comprises of old, end-of-life electronic appliances such as computers, laptops, TVs, DVD players, refrigerators, freezers, mobile phones, MP3 players, etc., which have been disposed of by their original users. E-waste contains many hazardous constituents that may negatively impact the environment and affect human health if not properly managed. Various organizations, bodies, and governments of many countries have adopted and/or developed environmentally sound options and strategies for E-waste management to tackle the ever growing threat of E-waste to the environment and human health.

6.4.1.1 E-waste Management

E-waste management is an emerging issue, driven by the rapidly increasing quantities of complex end-of-life electronic equipments. The international regulations, mainly under Basel convention, enforced a global ban on trans-border movements of E-waste but implementation of the same has its own difficulty due to lack of convergence in definition of E-waste followed by different governments, agencies and stake holders. Most E-waste in India ends up with slum based scrap dealers, who, in dangerous and hazardous working conditions take away resalable parts from this waste. The rest goes into landfills without proper treatment, posing serious ecological threat.

6.4.1.2 Impact of E-waste

Electronic wastes can cause widespread environmental damage due to the use of toxic materials in the manufacture of electronic goods. Hazardous materials such as lead, mercury and hexavalent chromium in one form or the other are present in such wastes primarily consisting of Cathode ray tubes (CRTs), Printed board assemblies,

Capacitors, Mercury switches and relays, Batteries, Liquid crystal displays (LCDs), Cartridges from photocopying machines, Selenium drums (photocopier) and Electrolytes. Although it is hardly known, e-waste contains toxic substances such as Lead and Cadmium in circuit boards; lead oxide and Cadmium in monitor Cathode Ray Tubes (CRTs); Mercury in switches and flat screen monitors; Cadmium in computer batteries; polychlorinated biphenyls (PCBs) in older capacitors and transformers; and brominated flame retardants on printed circuit boards, plastic casings, cables and polyvinyl chloride (PVC) cable insulation that releases highly toxic dioxins and furans when burned to retrieve Copper from the wires. All electronic equipments contain printed circuit boards which are hazardous because of their content of lead (in solder), brominated flame retardants (typically 5–10% by weight) and antimony oxide, which is also present as a flame retardant (typically 1–2% by weight) Land filling of e wastes can lead to the leaching of lead into the ground water. If the CRT is crushed and burned, it emits toxic fumes that are harmful to the environment.

These products contain several rechargeable battery types, all of which contain toxic substances that can contaminate the environment when burned in incinerators or disposed of in landfills. The cadmium from one mobile phone battery is enough to pollute 600 m^3 of water. The quantity of cadmium in landfill sites is significant, and considerable toxic contamination is caused by the inevitable medium and long-term effects of cadmium leaking into the surrounding soil. Because plastics are highly flammable, the printed wiring board and housings of electronic products contain brominated flame retardants, a number of which are clearly damaging to human health and the environment.

6.4.2 *Impacts of Informal Recycling in India*

The accrued electronic and electric waste in India is dismantled and sorted manually to fractions such as printed wiring boards, cathode ray tubes (CRT), cables, plastics, metals, condensers and other, nowadays invaluable materials like batteries. It is a livelihood for unorganized recyclers and due to lack of awareness; they are risking their health and the environment as well. The valuable fractions are processed to directly reusable components and to secondary raw materials in a variety of refining and conditioning processes. No sophisticated machinery or personal protective equipment is used for the extraction of different materials.

All the work is done by bare hands and only with the help of hammers and screwdrivers. Children and women are routinely involved in the operations. Waste components which does not have any resale or reuse value are openly burnt or disposed off in open dumps. Pollution problems associated with such backyard smelting using crude processes are resulting in fugitive emissions and slag containing heavy metals of health concern. CRT breaking operations result in injuries from cuts and acids used for removal of heavy metals and respiratory problems due to shredding and burning. They use strong acids to retrieve precious metals such as gold. Working in poorly ventilated enclosed areas without masks and technical expertise results in exposure to dangerous and slow poisoning chemicals. On a broader scale, analyzing the environmental and societal impacts of e-waste reveals a mosaic of benefits and costs. Proponents of e-waste recycling claim that greater employment, new access to raw materials and electronics, and improved infrastructure will result.

These will further boost the region's advance towards prosperity. Yet the reality is that the new wealth and benefits are unequally distributed, and the contribution of electronics to societal growth is sometimes illusory. Most e-waste "recycling" involve small enterprises that are numerous, widespread, and difficult to regulate. They take advantage of low labor costs due to high unemployment rates, internal migration of poor peasants, and the lack of protest or political mobilization by affected villagers who believe that e-wastes provide the only viable source of income or entry into modern development pathways. They are largely invisible to state scrutiny because they border on the informal economy and are therefore not included in official statistics.

6.4.3 Status of E-waste Management in India

As there is no separate collection of e-waste in India, there is no clear data on the quantity generated and disposed of each year and the resulting extent of environmental risk. The preferred practice to get rid of obsolete electronic items in India is to get them in exchange from retailers when purchasing a new item. The business sector is estimated to account for 78% of all installed computers in India. Obsolete computers from the business sector are sold by auctions. Sometimes educational institutes or charitable institutions receive old computers for reuse. It is estimated that the total number of obsolete personal computers emanating each year from business and individual households in India is around 1.38 million. According to a report of Confederation of Indian Industries (CII), the total waste generated by obsolete or broken down electronic and electrical equipment in India has been estimated to be 1,46,000 tons per year.

The results of a field survey conducted in the Chennai, a metropolitan city of India to assess the average usage and life of the personal computers (PCs), television (TV) and mobile phone showed that the average household usage of the PC ranges from 0.39–1.70 depending on the income class. In the case of TV it varied from 1.07–1.78 and for mobile phones it varied from 0.88–1.70. The low-income households use the PC for 5.94 years, TV for 8.16 years and the mobile phones for 2.34 years while, the upper income class uses the PC for 3.21 years, TV for 5.13 years and mobile phones for 1.63 years. Although the per-capita waste production in India is still relatively small, the total absolute volume of wastes generated will be huge. Further, it is growing at a faster rate. The growth rate of the mobile phones (80%) is very high compared to that of PC (20%) and TV (18%). The public awareness on e-wastes and the willingness of the public to pay for e-waste management as assessed during the study based on an organized questionnaire revealed that about 50% of the public are aware of environmental and health impacts of the electronic items. The willingness of public to pay for e-waste management ranges from 3.57–5.92% of the product cost for PC, 3.94–5.95% for TV and 3.4–5% for the mobile phones. Additionally considerable quantities of e-waste are reported to be imported.

However, no confirmed figures available on how substantial are these trans-boundary e-waste streams, as most of such trade in e-waste is camouflaged and conducted under the pretext of obtaining “reusable” equipment or “donations” from developed nations. The government trade data does not distinguish between imports of new and old computers and peripheral parts and so it is difficult to track what share of imports are used electronic goods.

Despite a wide range of environmental legislation in India there are no specific laws or guidelines for electronic waste or computer waste. As per the Hazardous Waste Rules (1989), e-waste is not treated as hazardous unless proved to have higher concentration of certain substances. Though PCBs and CRTs would always exceed these parameters, there are several grey areas that need to be addressed. Basel Convention has Waste electronic assemblies in A1180 and mirror entry in B1110, mainly on concerns of mercury, lead and cadmium. Electronic waste is included under List-A and List-B of Schedule-3 of the Hazardous Wastes (Management and Handling) Rules 1989 as amended in 2000 and 2003. The import of this waste therefore requires specific permission of the Ministry of Environment and Forests.

As the collection and re-cycling of electronic wastes is being done by the informal sector in the country at present, the Government has taken the following action/steps to enhance awareness about environmentally sound management of electronic waste:

- Several workshops on electronic waste management was organized by the Central Pollution Control Board (CPCB) in collaboration with Toxics Link and CII
- Action has been initiated by CPCB for rapid assessment of the E-Waste generated in major cities of the country.
- A national working group has been constituted for formulating a strategy for E-Waste management.
- A comprehensive technical guide on “Environmental Management for Information Technology Industry in India” has been published and circulated widely by the Department of Information Technology (DIT), Ministry of Communication and Information Technology.
- Demonstration projects have also been set up by the DIT at the Indian Telephone Industries for recovery of copper from Printed Circuit Boards. Although awareness and readiness for implementing improvements is increasing rapidly, the major obstacles to manage the e-wastes safely and effectively remain. These include:
  - The lack of reliable data that poses a challenge to policy makers wishing to design an e-waste management strategy and to an industry wishing to make rational investment decisions.
  - Only a fraction of the e waste (estimated 10%) finds its way to recyclers due to absence of an efficient take back scheme for consumers,
  - The lack of a safe e-waste recycling infrastructure in the formal sector and thus reliance on the capacities of the informal sector pose severe risks to the environment and human health.

– The existing e-waste recycling systems are purely business-driven that have come about without any government intervention. Any development in these e-waste sectors will have to be built on the existing set-up as the waste collection and pre-processing can be handled efficiently by the informal sector, at the same time offer numerous job opportunities.

6.4.4 *Waste Management Strategies*

The best option for dealing with e-wastes is to reduce the volume. Designers should ensure that the product is built for re-use, repair and/or upgradeability. Stress should be laid on use of less toxic, easily recoverable and recyclable materials which can be taken back for refurbishment, remanufacturing, disassembly and reuse. Recycling and reuse of material are the next level of potential options to reduce e-waste. Recovery of metals, plastic, glass and other materials reduces the magnitude of e-waste. These options have a potential to conserve the energy and keep the environment free of toxic material that would otherwise have been released.

An e-waste Policy is best created by those who understand the issues. So it is best for industry to initiate policy formation collectively, but with user involvement. Sustainability of e-waste management systems has to be ensured by improving the effectiveness of collection and recycling systems (for example, public–private-partnerships in setting up buy-back or drop-off centers) and by designing-in additional funding such as advance recycling fees.

6.4.4.1 E-waste Policy and Regulation

The Policy shall address all issues ranging from production and trade to final disposal, including technology transfers for the recycling of electronic waste. Clear regulatory instruments, adequate to control both legal and illegal exports and imports of e-wastes while ensuring their environmentally sound management should be in place. There is also a need to address the loop holes in the prevailing legal frame work to ensure that e-wastes from developed countries are not reaching the country for disposal. The Port and the Custom authorities in each country need to closely monitor these aspects. The regulations should prohibit the disposal of e-wastes in municipal landfills and encourage owners and generators of e-wastes to properly recycle the wastes. Manufactures of products must be made financially, physically and legally responsible for their products. Policies and regulations that cover Design for Environment (DfE) and better management of restricted substances may be implemented through measures such as

- Specific product take-back obligations for industry.
- Financial responsibility for actions and schemes.
- Greater attention to the role of new product design.

- Material and/or substance bans including stringent restrictions on certain substances.
- Greater scrutiny of cross-border movements of Electrical and Electronic Products and e-waste.
- Increasing public awareness by labeling products as "environmental hazard".

The key questions about the effectiveness of legislation would include:

- What is to be covered by the term electronic waste?
- Who pays for disposal?
- Is producer responsibility the answer?
- What would be the benefits of voluntary commitments?
- How can sufficient recovery of material be achieved to guarantee recycling firms a reliable and adequate flow of secondary material?

A complete national level inventory covering all the cities and all the sectors must be initiated. A public-private participatory forum (e-Waste Agency) of decision making and problem resolution in e-waste management must be developed. This could be a working group comprising of regulatory agencies, NGOs, industry associations and experts to keep pace with the temporal and spatial changes in structure and content of e-waste. This working group can be the feedback providing mechanism to the Government that will periodically review the existing rules, plans and strategies for e-waste management. Mandatory labeling of all computer monitors, television sets and other household/industrial electronic devices may be implemented for declaration of hazardous material contents with a view to identifying environmental hazards and ensuring proper material management and e-waste disposal. The efforts to improve the situation through regulations, though an important step, are usually only modestly effective because of the lack of enforcement. While there has been some progress made in this direction with the support of agencies such as GTZ, enforcement of regulations is often weak due to lack of resources and underdeveloped legal systems. Penalties for noncompliance and targets for collection or recycling are often used to ensure compliance.

6.4.5 Extended Producer Responsibility

Extended producer responsibility (***EPR***) is an environmental policy approach in which a producer's responsibility for a product is extended to the post consumer stage of the product's life cycle, including its final disposal. In principle, all the actors along the product chain share responsibility for the lifecycle environmental impacts of the whole product system. The greater the ability of the actor to influence the environmental impacts of the product system, the greater the share of responsibility for addressing those impacts should be. These actors are the consumers, the suppliers, and the product manufacturers. ***Consumers*** can affect the environmental impacts of products in a number of ways: via purchase choices (choosing environmentally friendly products), via maintenance and the environmentally conscious operation of products, and via careful disposal (for example, separated disposal of appliances for recycling). ***Suppliers*** may have a significant influence by providing manufacturers with environmentally friendly materials and components. ***Manufacturers*** can reduce the life-cycle environmental impacts of their products through their influence on product design, material choices, manufacturing processes, product delivery, and product system support. The system design needs to be such that there are checks and balances, especially to prevent free riders.

The goals of the product designer could include reducing toxicity, reducing energy use, streamlining product weight and materials, identifying opportunities for easier reuse among others. Manufacturers have to improve the design by: (i) the substitution of hazardous substances such as lead, mercury, cadmium, hexavalent chromium and certain brominated flame retardants; (ii) measures to facilitate identification and re-use of components and materials, particularly plastics; and (iii) measures to promote the use of recycled plastics in new products. Manufacturers should provide incentives to their customers for product return through a "buy back approach" whereby old electronic goods are collected and a discount could be given on new products purchased by the consumer. All vendors of electronic devices shall provide take-back and management services for their products at the end of life of those products. The old electronic products should then be sent back to be carefully dismantled for its parts to be either recycled or re-used, either in a separate recycling division at the manufacturing unit or in a common facility.

Collection systems are to be established so that e-waste is collected from the right places ensuring that this directly comes to the recycling unit. Collection can be accomplished through collection centers. Each electronic equipment manufacturer shall work cooperatively with collection centers to ensure implementation of a practical and feasible financing system. Collection centers may only ship wastes to dismantlers and recyclers that have authorization for handling, processing, refurbishment, and recycling while they meet environmentally sound management guidelines.

6.4.6 E-waste Recycling

Many discarded IT machines contain usable parts which could be salvaged and combined with other used equipment to create a working unit. It is labor intensive to remove, inspect and test components and then reassemble them into complete working machines. Institutional infrastructures, including e-waste collection, transportation, treatment, storage, recovery and disposal, need to be established at national and/or regional levels for the environmentally sound management of e-wastes. These facilities should be approved by the regulatory authorities, and if required, provided with appropriate incentives. Establishment of e-waste collection, exchange and recycling centers should be encouraged in partnership with governments, NGOs and manufacturers. Environmentally sound recycling of e-waste requires sophisticated technology and processes, which are not only very expensive, but also need specific skills and training for the operation.

Proper recycling of complex materials requires the expertise to recognize or determine the presence of hazardous or potentially hazardous constituents as well as desirable constituents (that is those with recoverable value), and then be able to apply the company's capabilities and process systems to properly recycle both of these streams. Appropriate air pollution control devices for the fugitive and point source emissions are required. Guidelines are to be developed for environmentally sound recycling of e-wastes.

6.4.7 Capacity Building, Training and Awareness Programs

The future of e-waste management depends not only on the effectiveness of local government, the operator of recycling services, but also on the attitude of citizens, and on the key role of manufactures and bulk consumers to shape and develop community participation. Lack of civic sense and awareness among city residents will be a major hurdle to keep e-waste out of municipal waste stream. Collaborative campaigns are required to sensitize the users and consumers should pay for recycling of electronic goods.

Consumers are to be informed of their role in the system through a labeling requirement for items. Consumers need to be educated to buy only necessary products that utilize some of the emerging technologies (that is lead-free, halogen-free, recycled plastics and from manufacturers or retailers that will "take-back" their product) to be identified through eco-labeling. Programs that raise the awareness on issues related to the environmentally sound management (ESM) and health and safety aspects of e-wastes in order to encourage better management practices should be implemented for different target groups. Technical guidelines for the ESM of e-wastes should be developed as soon as possible.

One of the biggest obstacles to the adoption of Green IT initiatives is the lack of awareness among employees, partners and senior administrators of the importance surrounding this issue and what each stakeholder can do to address them. Educating these stakeholders about emerging rules and regulations as well as making them aware of the growing business opportunities in Green IT can be critical drivers towards rapid adoption of Sustainable initiatives.

6.4.8 Summary

Solid waste management, which is already a mammoth task in India, is becoming more complicated by the invasion of e-waste, particularly computer waste. There exists an urgent need for a detailed assessment of the current and future scenario including quantification, characteristics, existing disposal practices, environmental impacts etc. Institutional infrastructures, including e-waste collection, transportation, treatment, storage, recovery and disposal need to be established at national and/or regional levels for the environmentally sound management of e-wastes. Establishment of e-waste collection, exchange and recycling centers should be encouraged in partnership with private entrepreneurs and manufacturers.

Model facilities employing environmentally sound technologies and methods for recycling and recovery are to be established. Criteria are to be developed for recovery and disposal of e-wastes. Policy level interventions should include development of e-waste regulation, control of import and export of e-wastes and facilitation in development of infrastructure. An effective take-back program providing incentives for producers to design products that are less wasteful, contain fewer toxic components, and are easier to disassemble, reuse, and recycle may help in reducing the wastes. It should set targets for collection and reuse/recycling, impose reporting requirements and include enforcement mechanisms and deposit/refund schemes to encourage consumers to return electronic devices for collection and reuse/recycling. End-of life management should be made a priority in the design of new electronic products.

The current practices of e-waste management in India suffer from a number of drawbacks like the difficulty in inventory maintenance, unhealthy conditions of informal recycling, inadequate legislation, poor awareness and reluctance on part of the corporate to address the critical issues. The consequences are that (i) toxic materials enter the waste stream with no special precautions to avoid the known adverse effects on the environment and human health and (ii) resources are wasted when economically valuable materials are dumped or unhealthy conditions are developed during the informal recycling. This section of the book chapter highlights the associated issues and strategies to address this emerging problem, in the light of initiatives in India.

6.5 Building the Business Case for Profit and Innovation in Indian IT Companies

The next step for IT companies is to build a business case that illustrates why it should pursue a set of principles/strategies that are well grounded in the principles of sustainability. These strategies could very well enable these companies to become market leaders while staying competitive in this sustainable environment. Firms are under increasing pressure from regulators, competitors, customers and community groups to implement innovative sustainable practices. Hence balancing economic and environmental performance to be green and competitive is a strategic issue worth pursuing. We now highlight some of the key strategies going forward

6.5.1 Engaging the Smart "Green" Grid

Green IT needs to be viewed from a holistic perspective that requires cross-disciplinary efforts. One area where IT will contribute involves the implementation of smarter "power" grids and in smart meters that can be employed in homes and commercial buildings. These smart meters and green grids will enable efficient monitoring, analysis and pricing of electrical usage. In India, where power production creates 60% of all emissions, a smart green grid can reduce power losses by nearly 15%. This could save over $ 13 billion a year while shrinking the carbon footprint at the same time.

Although Green Grids are useful metrics to help guide efficiencies, the data center industry, as a whole, is significantly behind other industries with regard to certification and regulation. Data centers consume about 30–100% more energy per square foot than office buildings partly due to the fact that the enormous amount of heat generated needs to be dissipated. In order to increase the energy efficiency, there has been an increased focus on the design of LEED rated data centers and use of ENERGY STAR rated data center servers worldwide. The first LEED rated data center was built by Netmagic, a leading Indian IT infrastructure provider. It has been estimated that this facility consumes, on an average, 20% less energy and 50% less water than conventional data centers[15]. Recently, the data center of HCL Infosystems, a leading Indian IT service vendor, was awarded LEED platinum rating[16], making it the first platinum rated LEED data center in the country.

6.5.2 Cloud Computing

One of the most important upcoming trends in the field of Information Technology is the concept of cloud computing. For long, managers in the IT world have been thinking of a way to increase data storage capacity or add computing capabilities on the fly without investing in new infrastructure, training new personnel or licensing new software. Cloud computing, which essentially is internet-based access to highly scalable pay-per-use IT capabilities is defined as a *potentially* cost effective model for provisioning processes, applications and services while making IT management simpler and increasing business responsiveness[17]. Cloud computing thus is a fundamental shift to an operational model in which applications don't live out their lives on a specific piece of hardware and in which resources are more flexibly deployed than was the historical norm. It's also a fundamental shift to a development and consumption model that replaces hard-wired, proprietary connections among software components and the consumers of those components with lightweight web services and web-based software access.

Cloud computing has lead to a convergence of technologies and trends that are making IT infrastructures and applications more dynamic, more modular and more consumable. Cloud computing provides the means to ramp up new services or reallocate computing resources rapidly, based on business needs. It means having the ability to run an application either on-premises or off-premises (or a combination of the two) based on cost, capacity requirements and other factors. Cloud shifts the costs for a business opportunity from CAPEX to OPEX, which allows finer control of

[15] http://www.netmagicsolutions.com/new-datacenter-launch.html

[16] http://www.hclinfosystems.in/news206.htm

[17] http://www-01.ibm.com/webcasts/podcasts/channels/gts/transcripts/cloud.pdf

expenditure and avoids costly asset acquisition and maintenance reducing the entry threshold barrier. Cloud computing is at an early stage, with a crew of large and small providers delivering a slew of cloud-based services, from full-blown applications to storage services to spam filtering. Broadly, the following are the types offered:

- **Infrastructure-as-a-Service (IaaS)**—IaaS lets the user or creator of the service provision processing, storage, networking, and other computing resources on which they can then run operating systems and applications. The deployer of the service does not manage or control the underlying compute infrastructure and is responsible only for configuring and maintaining the software. Example—IBM Cloud
- **Platform-as-a-Service (PaaS)**—PaaS provides computational resources via a *platform* upon which applications and services can be developed and hosted. With PaaS, one can build entire applications from scratch on a leased platform without having to buy and maintain the underlying hardware and software. Example—Amazon web services.
- **Software-as-a-Service (SaaS)**—Sometimes referred to as *Service* or *Application* clouds, offer implementations of specific business functions and business processes as direct consumables for end users. Example—SALESFORCE.COM CRM

Cloud computing comes with a lot of benefits at all levels of an organization where it is deployed. At the macro level, cloud computing catalyzes achieving economies of scale by centralizing compute power and democratizing access to all resources At the CIO level, cloud computing helps shift the mindset to commoditize computing power, therefore drive efficiencies through greater utilization rates which allows systems to scale up or down due to load fluctuations and not idle when not required. At the data center level, cloud computing's drive towards consolidation paves the way for new standards for energy efficiency, as discussed in the previous section. At the R&D level, cloud computing creates incentives for software engineers to code more efficient and re-usable applications so that they can be hosted.

In order to understand the environmental benefits of cloud computing, one has to understand the following three independent factors[18] that contribute to cutting a company's carbon footprint, either directly or indirectly:

Dynamic Provisioning A methodology in which computing resources are provided on need basis so that actual demand and server capacities allocated are more closely matched. Infrastructure planning by firms is typically conducted with a conservative, "just in case" mindset that results in capacity allocation that is not aligned with actual demand. On the other hand cloud providers can manage capacity much more diligently and accurately by forecasting their client's demands. Dedicated personnel monitor and predict demand and continually adjust capacity thus negating inefficiencies in resource allocation. This leads to lesser computational needs, lesser

[18] Cloud computing and sustainability—The environmental benefits of moving to the Cloud—Accenture consulting paper 2010

requirement of cooling and maintenance of the servers, leading to energy savings and reduction in the firm's carbon footprint.

Multi-Tenancy Just as multiple tenants in an apartment building use less power overall than the same number of people owning their own homes, so do the multiple tenants of a cloud-provided infrastructure reduce their overall energy use and associated carbon emissions. The cloud architecture allows providers to simultaneously serve multiple companies on the same server infrastructure. Disparate demand patterns from numerous companies flatten overall demand peaks and make fluctuations more predictable. The ratio between peak and average loads becomes smaller, and that in turn reduces the need for extra infrastructure. Major cloud providers are able to serve millions of users at thousands of companies simultaneously on one massive shared infrastructure. By operating multi-tenant environments, cloud providers can reduce overhead for on-boarding and managing individual organizations and users.

Server Utilization Cloud computing can dramatically improve the utilization of servers and consequently help in saving energy and reducing a company's carbon footprint. It has been found that while the utilization of on-premise servers typically ranges between 5–10%, utilization of servers on the cloud can be as high as 40–70%[19], thus increasing the numbers of users served per machine sharply.

According to a study done by a global consulting firm—Zinnov Management Consulting, Cloud Computing market in India, which currently stands at $ 110 million may go up to $ 1 billion by 2015. Software-as-a-service (SaaS) in India is likely to reach a mark of $ 650 million by 2015, while Platform-as-a-service (PaaS) and Infrastructure-as-a-service (IaaS) markets cumulatively would touch 434 million each by then. The growth of IT spending in India is one of the fastest in the world and is expected to be higher than China soon. India is the world's fastest growing mobile market with over 20 million subscribers added every month. Companies have invested heavily in services such as 3G for enhanced data utilization. The emergence of Cloud technology is also linked to the rise in Telecom sector. Indian customers are ideal for cloud offerings, as these offerings enable cost savings, less maintenance worries and better operability.

So how does Cloud help in Sustainability? The Information and Communication Technology (ICT) contributes to as much as 15% of total global carbon emissions (or 7.8 billion tons of CO2 equivalents per year). Broad adoption of cloud computing can stimulate innovation and accelerate the deployment of these enabled solutions. Consequently, cloud computing may have a major impact on global carbon emissions through indirect benefits in addition to the direct savings from replacement of on-premise infrastructure. An average data center's Power Usage Effectiveness (PUE) is typically around 2.0 or 50%. Massive data centers built by Google, Amazon and Microsoft typically operate on a PUE of 1.2 or 1.3—Reason—Large scale application of water and air cooling techniques.

[19] Silicon Valley Leadership Group. Data Center Energy Forecast. 2008. IDC. The Economics of Virtualization: Moving Toward an Application-Based Cost Model—2009

It will be some years before people realize the true potential of the Cloud. In conclusion, though Cloud Computing is in its nascent stage, it holds a lot of promise, not just for companies to reduce their operating expenses, but also for the environment. Increasing the density of hardware used and introduction of cloud computing in the data management space have been identified as the future trends of data center greening. It has been estimated that enterprises can save $ 4 billion every year just by following more efficient practices in managing their data centers[20]. Through better management of assets, more accountable management and setting clear goals for reducing energy costs and carbon emissions, most companies can double IT efficiency and halt the growth of greenhouse gas emissions.[21]

6.6 Going Forward

Environmental sustainability is one of the most important global challenges that we have to deal with in this century. Organizations that want to remain competitive need to play a critical role in promoting environmental sustainability and mitigating climate changes. Both financial and environmental gains can be made by staying in touch with both customers and clients. IT can play a huge role in this journey by offering innovative solutions such as using electronic communication (decrease printing costs) and using voice, video and on-screen collaboration for connecting geographically dispersed employees.

Companies such as Wipro and TCS are using IT to lower its carbon footprint through several means. They include virtual meetings using e-collaboration tools such as webinar, videoconferencing and telephony; streamlining governance using automated reporting and business intelligence dashboard and using technology to obtain real time data about products and processes as well as real time visibility into this data using RFID and GPS technology. For example, the mKRISHI project at TCS (www.tcs.com) uses a sensor network of weather and soil sensors, camera phone, cellular networks and GPS technologies to provide customized environmental information to grape, cotton and soybean farmers in server rural India villages.

From a sustainable perspective, IT has enabled firms to standardize, monitor, capture and utilize data that helps obtain energy efficiencies. TCS has enforced a reduce, reuse and recycle policy for the lifecycle of all of its products from procurement to disposal. Donating equipment to charities and NGOs along with recycling all its equipment in a proper manner are key goals that are being practiced by the company. Using online forms and streamlining both their business and workflow processes have yielded a significant reduction in their paper consumption. Additionally,

---

[20] Cloud computing and sustainability—The environmental benefits of moving to the Cloud—Accenture consulting paper 2010

[21] http://www.mckinsey.com/clientservice/ccsi/pdf/Data_Centers.pdf

measures to enforce double-sided printing (duplex), shared printing services and recycling printer cartridges have all led to significant reduction in paper consumption and printer cartridges.

References

Garrod B (1997) Business strategies, globalization and environment. In Global and Environ OECD Paris, 269–314

Gladwin T, Kennelly J, Krause T (1995) Shifting paradigms for sustainable development: Implications for management theory and research. Acad of Manage Rev 6(2):874–907

Hart SL, Milstein MB (2003) Creating sustainable value. Acad of Manage Exec 17(2):56–67

Mines C (2008) The dawn of Green IT services: A market overview of sustainability consulting for IT organizations. Forrester Res Rep, Cambridge

Molla A Cooper VA, Pittayachawan S (2009) IT and eco-sustainability: Dev and validating a Green IT Readiness model. ICIS (2009) Proceedings, 1–17

Narasimhan PLL, Raghuraman K (2010) http://www.wipro.com/greenit/pdfs/Five%20Steps%20to%20Green%20Data%20Centres.pdf

Steadman ME, Zimmerer TW, Green RF (1995) Pressures from stakeholders hit Japanese companies. Long Range Plann 28(6):29–37

Tarafdar M, Modi S (2010) Adoption of Sustainability in IT services: The Role of IT Service Providers. Americas Conf on Inf Syst Peru

Thibodeau P (2007) Gartner's Top 10 strategic technologies for (2008). Computerworld October 9

Watson RT, Boudreau M, Chen AJ (2010) Information systems and environmentally sustainable development: Energy informatics and new directions for the IS community. MIS Quarterly 34(1):23–38

Chapter 7
Development of a LEED Certified Building—A Case Study

Rex M. Holmlin

Abstract The design and construction of Miller Hall, a 166,000 square foot $75 million dollar LEED Gold certified classroom and office building is the first link in a green supply chain for undergraduate, graduate and continuing corporate education. The building also serves as a platform for a wide variety of faculty research and staff and student activities focusing on sustainability.

7.1 Project Background and Overview

In August 2009 the Mason School of Business at the College of William & Mary began classes in a new classroom and office building. Miller Hall, a LEED Gold certified 166,000 SF building, may be viewed as one link in a green supply chain for business education for undergraduate, graduate and corporate continuing education. The new building serves as a platform for a wide variety of sustainable research and educational activities within the Mason community. This chapter describes the design and construction process with emphasis on those activities involved with the LEED certification process.

Prior to August 2009, the Mason School of Business was housed in four locations. Undergraduate and graduate classes were taught in two different buildings with most support functions in those buildings; however, offices for the School's development staff and some enterprise support staff were in two additional off-campus locations. This was inefficient from a management perspective and was not conducive to building a sense of community within the various members of the Mason School. Following a 5 year period of feasibility studies, architectural programming and fundraising a design and construction effort began in late 2005.

The new building project was to be executed through a public-private partnership. The public entity partner was the College of William & Mary and the private entity partner was the Business School Foundation of the College of William & Mary.

The College of William & Mary is part of the state university system of the Commonwealth of Virginia. Originally founded in 1693 by royal charter of King

R. M. Holmlin (✉)
Mason School of Business, The College of William and Mary,
101 Ukrop Way, Williamsburg, VA 23185, USA
e-mail: rex.holmlin@mason.wm.edu

T. Boone et al. (eds.), *Sustainable Supply Chains*,
International Series in Operations Research & Management Science 174,
DOI 10.1007/978-1-4419-6105-1_7,

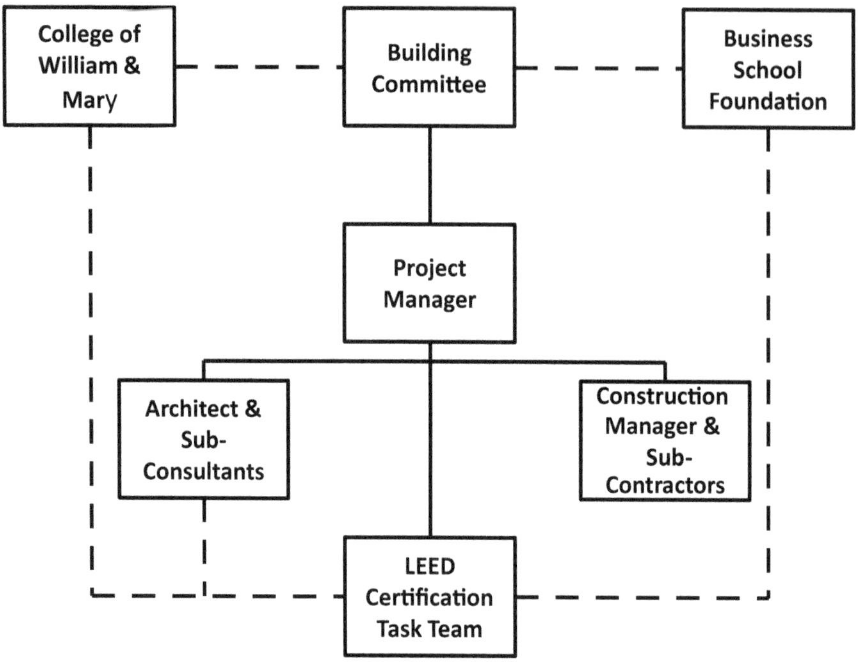

Fig. 7.1 Project organization chart

William III and Queen Mary II, the College of William & Mary is the second oldest institution of higher learning in the United States. At the time of the Miller Hall project there were 570 faculty and 7,700 students at the College of William & Mary. The Mason School of Business School had approximately 110 faculty and staff, 300 graduate students and approximately 1,000 undergraduates.

The responsibility for the development of the new building was assigned to the Business School Foundation through a Memorandum of Understanding between the College and the Business School Foundation. Management oversight of the project was vested in a nine member building committee with five members from the Business School Foundation and four members from the College of William & Mary. The Building Committee was co-chaired by the Dean of the Business School and the Assistant Vice President for Administration from the College. The responsibility for the day-to-day management of the project rested with a project manager retained by the Business School Foundation. Figure 7.1 shows the organization of the project.

Based on feasibility and programming studies the new building was to be approximately 166,000 SF in size with an approximately 61,000 SF 'U' shaped footprint. The building was to have three stories and a partial basement. To be compatible with the older, or ancient campus at the College, the building exterior was to be Georgian style although the Building Committee directed that the building interior was to convey a sense of a twenty first century business school.

The project budget was $75 million. Table 7.1 shows the major cost categories of the project's budget. Funding for the project was provided through the public-

Table 7.1 Project budget

Budget category	Budget ($)	Cost/SF ($)	As percentage of budget
Architect-engineer fees	7,465,881	44.98	9.95
Pre-construction services	330,000	1.99	0.44
Construction (GMP)	52,107,701	313.90	69.48
Furniture, fixtures & equipment	3,439,800	20.72	4.59
IT/AV & security	3,531,978	21.28	4.71
Project inspection	420,000	2.53	0.56
Other costs	4,060,000	24.46	5.41
Owner's contingency	3,644,640	21.96	4.86
Total	*75,000,000*	*451.81*	*100.00*

private partnership. Two-thirds of the project's budget ($50MM) was provided by the private entity portion of the partnership and one third ($25MM) was provided from the public entity of the partnership. $25MM of the Foundation's funding was to be donations and $25MM was to be raised through issuance of bonds. All of the College's funding was to be raised through the issuance of a separate $25MM of bonds. Any project cost overruns were to be paid for by the private entity portion of the partnership. Not surprisingly, project cost management was extremely important to the Business School Foundation.

The project used a construction manager at risk with a guaranteed maximum price form of project delivery.[1] This allowed the project team to have the benefits of construction expertise during the design phases of the project. A pre-construction services contract for $330,000 was executed with the Whiting-Turner Contracting Company following a selection process that included submission of qualifications, interviews and review of references. Key duties of the Construction Manager during the design phase included cost estimating, verification of scope-budget match and constructability review of the design documents.

The project had two architects. The Architect of Record, responsible for the construction documents, was Mosley Architects headquartered in Richmond, Virginia. The design architect, responsible for the design concept, was Robert AM Stern (RAMSA) of New York city. From a contractual perspective the design architect was a sub-consultant to the architect-of-record. Design began in the spring of 2006 and included a number of benchmarking trips to other business schools to look at recent design concepts for business schools particularly with regard to classrooms and teaching walls.

Architectural programming called for 14 classrooms, 2 seminar rooms, a financial markets classroom, 28 team rooms for student and project teams, a café and a 4,000 SF multi-purpose room.

[1] The Construction Manager at Risk with a Guaranteed Maximum Price form of project delivery allows an Owner to select a construction manager on the basis of qualifications and price; typically early in the project the design is not completed so the price selection is based on pre-construction services. When the design is more advanced, but still not complete, the construction manager provides a guaranteed maximum price (GMP) which provides a cost ceiling and a degree of cost protection for the Owner.

The site chosen for the project was a parking lot located on Jamestown Road at the western gateway to the William & Mary campus. The new building would front on Ukrop Drive and have a building axis parallel to the longitudinal axis of the sunken garden of the historic campus of the College. The site selected for Miller Hall site was originally developed by a project done by the Works Progress Administration (WPA) during the 1930's. That project included a stable and a riding ring for students. Over the years the stable became the Center for Archeological Research (CAR) and the riding ring became a parking lot for day students and staff. As part of the Miller Hall project, the parking lot and CAR building would be demolished.

Project cost management was an important aspect of the project management effort and, as noted earlier, was of great interest to the Business School Foundation. As design began, the project team developed a building cost estimate using an analogous estimating technique.[2] This allowed the project team to identify more accurately the square footage of the new building that was affordable and to compare this budget feasible size to the proposed architectural program. The draft architectural program was approximately 20,000 SF larger than the budget feasible size of the building based on the analogous cost estimate. This allowed the project team to trim back the architectural program which was the basis for design. The data for the similar projects used in the analogous estimate came from the architects and the construction manager. This data included not only total costs for the comparable projects, but data for key components of the building including electrical and mechanical systems. This information, in the standard Construction Specifications Institute (CSI) 16 Division[3] format, allowed the project team to develop design to cost targets for the architects and sub-consultant engineers. As the design progressed, the project team also used the Construction Industry Institute's (CII) Project Definition Rating Index (PDRI) to help evaluate the completeness of the project scope definition (Construction Industry Institute 1999). Other cost management tools included value engineering studies conducted at the end of schematic design and reconciliation of cost estimates and both the schematic and design development phases of design.

To aid in constructability and in cost management the project team identified several key sub-contractors based on the importance of their work in the project. For instance, because there were more than 500,000+ bricks on the building exterior and the Georgian architecture was a critical architectural design feature, the brick mason was identified as a key trade and included as a design-assist sub-contractor. Other sub-contractors selected to participate in the design-assist program included the mechanical, electrical and structural steel erector. These sub-contractors were selected on a qualifications and cost basis and brought on early to work with the architects and engineering sub-consultants. These sub-contractors provided input to the design team on issues of constructability and cost. This paid significant benefits

---

[2] Analogous estimating uses similar or analogous projects, as a basis for developing a cost estimate. The cost of each comparable project is adjusted for location and date of construction to assist in determining a cost for the subject project.

[3] The CSI 16 Division format was standard in the construction industry until 2004 and included 16 categories such as sitework, steel, woodwork, mechanical and electrical components of work.

during construction in not only ease of construction and cost, but also construction quality and schedule.

The target completion date for the project called for classes to begin in August of 2009. Including time required for commissioning of the building's mechanical systems, move in of faculty and staff as well as spin-up of the new building a substantial completion in early June of 2009 was required. Based on a projected 26 month construction schedule, the project team opted for a fast-track approach to the overall schedule which allowed for a March 30, 2007 groundbreaking. However, as is common in fast-track projects, the design was not complete at the time of groundbreaking. While the site, building foundation and structural steel design was complete, interior design was still underway when construction began.

The construction team made use of mock-ups including exterior wall mock-ups as well as classroom mock-ups and extensive use of "first-work" to establish quality standards for follow-on work. The project had several types of classrooms including traditional tiered classrooms and a new cluster classroom design that allowed students to move easily from a lecture mode to being "clustered" around oval shaped table to work in groups for more experiential learning. As part of the design and construction process, classroom mock-ups were built and faculty participated in final adjustments of tier heights, sight-lines and the shape of tables in the classrooms.

The project achieved substantial completion on schedule in June 2009 and was slightly under budget. Faculty and staff began to move into the building the week after the July fourth holiday. Within a week of move-in, the first corporate continuing education courses were being held as part of the building's "spin-up" process. Undergraduate and graduate classes began in August 2009.

7.2 Green Building History

At the time of the Miller Hall project, the US Green Building Council's LEED program was 5 years old; however, the roots of green building and sustainability stretched back three decades to the mid-1970's. Building projects generate impacts through siting, their design as well as their construction, operation and maintenance and eventual renovation and demolition. These impacts are generated by consumption of construction and other materials, use of natural resources as well as energy and water. Typical environmental effects include construction and operation wastes, air and water pollution, poor indoor air quality, degradation of surface water through stormwater runoff, noise and heat island effects. Green buildings are designed to have reduced environmental impact by (1) a more efficient use of resources including energy and water, (2) safeguarding the health of occupants and creating a work environment that allows for greater productivity and (3) reducing degradation of the environment through reduced waste and pollution (Environmental Protection Agency (2010)).

By the late 1980's and early 1990's interest in sustainability had gained significant momentum. Although conversations around energy efficiency had begun in the

1970's, in 1989 the American Institute of Architects (AIA) formed the Committee on the Environment (Gould 2007). Three years later, with funding from the EPA, the AIA published the "Environmental Resource Guide". The same year the Department of Energy and EPA launched the Energy Star program.

Energy Star, a voluntary labeling program, was designed to identify and promote energy efficient products with a goal of reducing emissions of green house gases. Since its' inception the program has expanded to provide labeling on more than 60 product categories. Technical information and tools to assist in product selection are provided to consumers based on more than 20,000 partnerships with both private and public sector organizations. It is estimated that the Energy Star program saved consumers, business and other organizations approximately $18 billion in 2010 alone (Energy Star (2010)).

The US Green Building Council was formed in 1993; 5 years later they launched a pilot version of the Leadership in Energy and Environmental Design (LEED) program. By 2000 LEED was no longer a pilot program. It had become an internationally recognized system that allows building owners to measure approaches to green building design, construction, operation and housekeeping. LEED is a suite of rating systems for new construction, renovation, schools, healthcare, residential and retail development (EPA (2010)).

The LEED program has several certification levels including certified, silver, gold and platinum. Each certification level requires a different number of points on the LEED rating index. LEED Certified buildings require a minimum of 26 points while silver certified buildings require a minimum of 33 points. Gold buildings require at least 39 points and platinum certified buildings require a minimum of 52 points. The Miller Hall project sought certification under LEED's new construction standards (NC 2.2) (Miller Hall Project Records (2010)).

7.3 Miller Hall Sustainability and LEED Certification

Early guidance to the project team from the Building Committee directed that Miller Hall should be a '100 year building'. Building codes do not define what the standards for a 100 year building are; however, with this guidance the Building Committee conveyed its' interest in both sustainability and long term value to the project team.

As design got underway in early 2006, the project team was reviewing options for a sustainable building based on the guidance from the Building Committee. In addition to potential energy and cost savings, the project team was also studying information on health and productivity gains in sustainable buildings (Fisk (2000) and Rosenberg and Neir (2003)).

At the February 22, 2006 meeting of the Building Committee, the project team recommended to the Committee that the project pursue LEED certification with a goal of achieving Silver LEED certification (Miller Hall Project Records (2006a, c)). This recommendation was approved by the Building Committee and the project team moved forward with LEED planning and application for certification.

To implement the goal of achieving LEED certification, the Project Team established a LEED task team with membership from the Architect-Engineer and key sub-consultants, the Construction Manager, operations and maintenance staff from the College of William & Mary and the Foundation's project manager (see Fig. 7.1). On March 21, 2006 a "kick-off" meeting for the LEED certification efforts was held with key members of the project and sub-consultant team and in early April, 2006, the project team briefed the College's Landscape, Environment and Energy Committee (LEE) regarding the project's goal of LEED certification (Miller Hall Project Records (2006b)). Because the Building Committee had decided early in the design process to pursue LEED certification, the project team was able to integrate the LEED process with the design process. This paid significant benefits in terms of schedule, cost of design and quality of sustainable design for the project.

Interest in sustainability was also growing in the wider Mason Community. Several faculty were pursuing research in the area of sustainability and material on sustainability was being incorporated into academic course materials. Importantly, as the early design efforts of the project team were underway, the undergraduate business program was working on a report "Curricular Issues related to Environmental Initiatives for Sustainability" (Geary and Adkins 2006). This report looked at curriculum, case study competitions and other sustainable activities across the William & Mary campus. It also urged consideration of LEED certification for the new building then under design.

LEED certification through the U.S. Green Building Council (USGBC) is a two step process. After design is complete a design application is submitted and reviewed by the Council. Once construction is complete a similar application covering construction phase activities is submitted and reviewed by the Council. Based on the total number of points awarded from both reviews, the USGBC may issue a LEED certification.

The US Green Building Council LEED process has identified six broad categories of sustainability that a project may receive points in. These include (1) sustainable site, (2) water efficiency, (3) energy and atmosphere, (4) materials and resources, (5) indoor air quality and (6) innovation and the design process. Within each of the main categories there are sub-categories and both required and potential credits that have been identified. As the design continued and the LEED certification process started, the LEED task team conducted a series of reviews looking at the various categories where sustainability could be incorporated into the design. Creativity and "out-of-the-box" thinking was encouraged at these sessions. As potential sustainability features were selected for incorporation or tagged for further investigation the key constraints on the LEED team were the building's Georgian design and operation and maintenance considerations. For instance items such as a green roof were neither Georgian and, from the perspective of the College's maintenance team, represented a high potential for future problems. Key sustainable design features that were incorporated into the building included:

Sustainable Site Development. In this category the selected site was near ten community services and was also within 1/4 mile of two bus lines. The site design also provided bicycle storage, changing rooms for those who bicycled to work and parking spaces for low emitting vehicles. The site design also provided for a stormwater management design that did not increase stormwater rate or volume of discharge when compared to pre-development conditions and the design limited heat island effect by providing for nearly 58% of non-roof paved areas to be shaded within 5 years.

Water Efficiency. The project's design incorporated an underground cistern that captures stormwater runoff from the roof as well as condensation from mechanical equipment. By using dual flush water closets, low flow urinals and ultra-low flow lavatories projected water usage was reduced to less than 33% of a calculated baseline. and uses it for irrigation including water efficient landscaping and water use reduction.

Energy & Atmosphere. The design focused on attempting to optimize energy performance. The building's lighting and mechanical systems achieving a forecasted 20% energy savings over ASHRAE standard 90.1 (i.e. code) compliant building. The project also purchased green power credits.

Materials & Resources. These design features include diversion of construction waste, recycled content, use of regional materials and certified wood. The building's design includes recycling stations on each floor to collect paper, plastics and aluminum and glass containers. 1,984 tons of construction waste (83.4% of the total) was diverted from construction waste landfills and recycled. 20% of the building's material content was recycled materials and nearly 43% of the value of materials used in construction were manufactured within 500 miles of the project site. 60% of the wood based materials in the project were harvested from Forest Stewardship Council (FSC) certified forests.

Indoor Environmental Quality. Key design features included use of CO_2 monitors, indoor air quality control during both construction and pre-occupancy, use of low emitting materials including paints, adhesives, sealants, carpets and composite wood subsystems, controllability of systems for both lighting and thermal comfort as well as thermal comfort design and verification. Other sustainable design features include occupancy sensors which cause building lighting to automatically shut off when no one is in a room. Additionally, procedures were established that provide for the building to participate in the College's Green Housekeeping initiative aimed at reducing exposure to cleaning chemicals and practices that are harmful to people and the environment.

Innovation & Design Process. Some of the design features in this category including a tree protection plan for the site that allowed for twice as many trees to be saved as had been planned for in the 100% schematic design phase drawings. The building also has a green building education program that has information plaques about sustainable building features throughout the building. The project also used low emitting furnishings and had LEED accredited professionals on the project team.

As both the design and LEED certification efforts proceeded, it became apparent that the project might have sufficient points to achieve Gold level certification which is a higher level than the target of Silver certification. Building tours during construction

Table 7.2 Summary of USGBC points awarded by category

USGBC LEED-NC 2.2 category	Points earned by the project	Possible points in the category
Sustainable site	10	14
Water efficiency	4	5
Energy & atmosphere	6	17
Materials & resources	7	13
Indoor air quality	11	15
Innovation & design process	4	5
Total	*42*	*69*

also showed a high interest in sustainability in all parts of the Mason Community. While achieving Silver certification remained the public goal, there was so much interest in sustainability in general, and in achieving Gold certification for the building, that not achieving Gold level certification would have been a disappointment to the Project Team, Building Committee and wider Mason community.

In April, 2010 the School was notified that the project had been awarded LEED Gold certification. Table 7.2 details the points that the project was awarded based on the US Green Building Councils evaluation of application materials (Miller Hall Project Records (2010)).

7.4 Lessons Learned

Five key lessons emerge from this project's experience with LEED certification:

a. It is critical to have the support of senior leadership. The strong support the project team received from the Building Committee was instrumental in achieving LEED certification. As previously noted it also allowed LEED planning to be fully integrated into the design and construction process and opposed to being an "add on" process.
b. A designated LEED task team provided the organizational focus and responsibility within the larger project team for this key task.
c. The LEED certification effort assisted the project team in maintaining a "life cycle" approach to the project by helping integrate longer term operational and housekeeping considerations into project design and construction.
d. The initial submittals for both design and construction to the US Green Building Council did not receive as many points as the project team felt were warranted. The project team took advantage of the appeals process and submitted additional information to the USGBC to support the project team's request for certification points. This was critical in gaining the 42 points the project received which resulted in Gold Certification for the project.
e. As construction continued and the project team gave tours to members of both the Mason School community and the public, it became clear the sustainable

building and LEED certification were important to a wide spectrum of the Mason community and the public. LEED certification became tangible evidence of the project's commitment to sustainability and environmental values.

7.5 Conclusion

The construction, and LEED certification, of Miller Hall provided a key link in a sustainable educational supply chain. First, the building provides tangible evidence of the Mason community's commitment to, and campus leadership in, sustainability. The Building has also helped provide a platform for follow-on sustainability efforts such as the Do One Thing (DOT) sustainability campaign and research and teaching by faculty as well as broader participation by staff and students in sustainability efforts.

References

Construction Industry Institute (1999) Project definition rating index—building projects, implementation resource 155–2. Construction Industry Institute, Austin

Environmental Protection Agency (December 22, 2010) Green Buildings. http://www.epa.gov/greenbuilding/pubs/about.htm. Accessed 24 June 2011

Fisk W (2000) Health and productivity gains from better indoor environments and their relationship with building energy efficiency. Annu Rev Energy Environ 25:537–566

Gould K (2007) AIA/COTE: A history within a movement. http://www.aia.org/practicng/groups/kc/AIAS077347. Accessed June

Geary W, Adkins C (2006) Curricular issues related to environmental initiatives for sustainability. Williamsburg: Mason School of Business, College of William and Mary

Miller Hall Project Records (2006a) February 22, 2006 Meeting Minutes Miller Hall Building Committee

Miller Hall Project Records (2006b) 21 March 2006 Meeting Minutes LEED Task team "kick off" Meeting

Miller Hall Project Records (2006c) Notes for presentation to 6 April 2006 meeting of college of William and Mary Landscaping, Energy and Environmental Committee

Miller Hall Project Records (2010) LEED-NC Construction Review Application, 2 April 2010

Rosenberg L, Neir A (2003) Health benefits of green buildings. Paper presented at the Delaware Valley Green Building Council. Philadelphia

Chapter 8
A Primer on E-Waste

Tonya Boone and Ram Ganeshan

Abstract This Chapter introduces the concept of e-waste. The chapter highlights how countries are regulating e-waste and how firms that sell electronic and electrical equipment are tackling e-waste with supply chain initiatives.

8.1 What is E-Waste?

Electronic waste or e-waste (often referred to as Waste Electrical and Electronic Equipment (WEEE)) is the term used to describe old, discarded, surplus, obsolete, or broken electrical or electronic devices (any device that runs on electricity or batteries) regardless of whether an appliance is still functional or not. It includes computers, phones, consumer electronics, and household appliances that have reached end of their useful life. Figure 8.1 gives a breakdown of the contents of e-waste[1].

The fast pace of technological change, shortening life spans of devices, high rates of obsolescence (or replacement) in developed countries, and the growing demand for electrical and electronic equipment (EEE) devices in the developing countries have all contributed to the rapid growth of e-waste.

In 2009, approximately 53 million metric tons of e-waste was generated worldwide and about 13% of that weight was recycled[2]. Estimating the quantity of e-waste can be complex for several reasons. First, different products have different life spans and each user will use the product differently making it difficult to predict when a product will become eligible as "waste." Second, once the product reaches the end

All web sites referenced in this Chapter were accessed 15th August 2010. We would like to thank Karen Marshall and Shoumya Prodhan for helping research issues on e-waste.

[1] EMPA Swiss Federal Laboratories for Materials Testing and Research (definition according to the European Union) http://maps.grida.no/go/graphic/what_is_e_waste.

[2] http://www.bbc.co.uk/news/world-10851645.

T. Boone (✉) · R. Ganeshan
Mason School of Business, College of William and Mary, 23185 Williamsburg, VA, USA
e-mail: Tonya.Boone@mason.wm.edu

R. Ganeshan
e-mail: Ram.ganeshan@mason.wm.edu

T. Boone et al. (eds.), *Sustainable Supply Chains*,
International Series in Operations Research & Management Science 174,
DOI 10.1007/978-1-4419-6105-1_8,

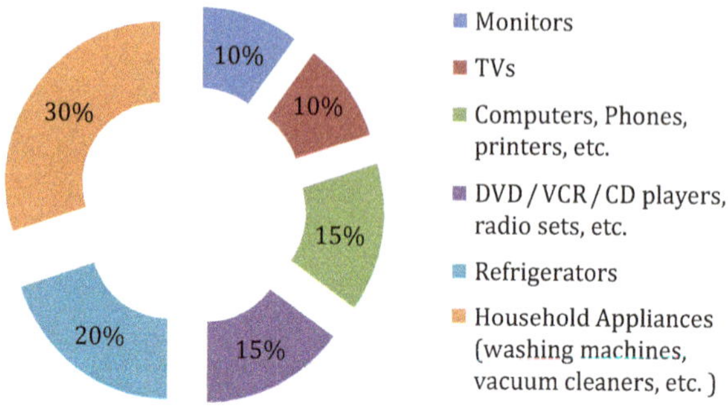

Fig. 8.1 Whats in E-waste? (Source: EMPA 2005)

of its primary life span (or the 1st user), it officially enters its "end of life" and its management by the user is also uncertain. At this stage, it can be recycled, landfilled, or (some portions) incinerated as waste. The product, however, can also be reused for a "second life" or can simply be put in storage. Those items that are either reused or in storage will eventually be recycled, landfilled, or incinerated at a future date.

The estimation of e-waste can be thought of in terms of flows. Historical sales figures give an estimate of the potential amount of recoverable devices in use or storage. Landfilled and recycled amounts give a picture of how much of that potential amount is recycled. Focused policies and strategies can optimize the "recovery rates" to close the loop in the supply chain.

In 2005, the US EPA estimates that 304 million electronic devices (computers, TVs, VCRs, cell phones, and monitors) were discarded from US households. The Consumer Electronics Association (CEA) estimates that of the devices that were discarded, about two-thirds were still functional.

Used or unwanted electronics in 2005 amounted to approximately 1.9–2.2 million tons. Of that, about 1.5–1.8 million tons was landfilled, and only a fraction—345,000–379,000 tons—was recycled.

Figure 8.2 shows the sales for popular electronic products in the US from 1990 through 2005[3] along with the "end of life" management for the year 2005 for these products. In 2005, for example, 3.56 million tons of electronic equipment was sold (and 1.5–1.8 tons entered the waste stream) and 636,000 tons of it was desktops. In 2005, 608,500 tons of desktops became eligible for "end of life" of which 349,000 tons were either reused or put into storage[4], 186,500 tons found their way to a landfill, 67,800 tons were recycled, and 5,300 tons were incinerated.

---

[3] Data is from the US Environmental Protection Agency: http://www.epa.gov/osw/conserve/materials/ecycling/manage.htm.

[4] Note that devices that become eligible for end of life were actually sold in previous years—so while current year sales is one point of reference, previous year sales are a better benchmark. Dell and Apple, for example, use sales 7-years prior to compute recycle rates. These devices that go into storage or reused are technically not "retired" but will find its way to the waste stream in the future either from storage or after it is reused.

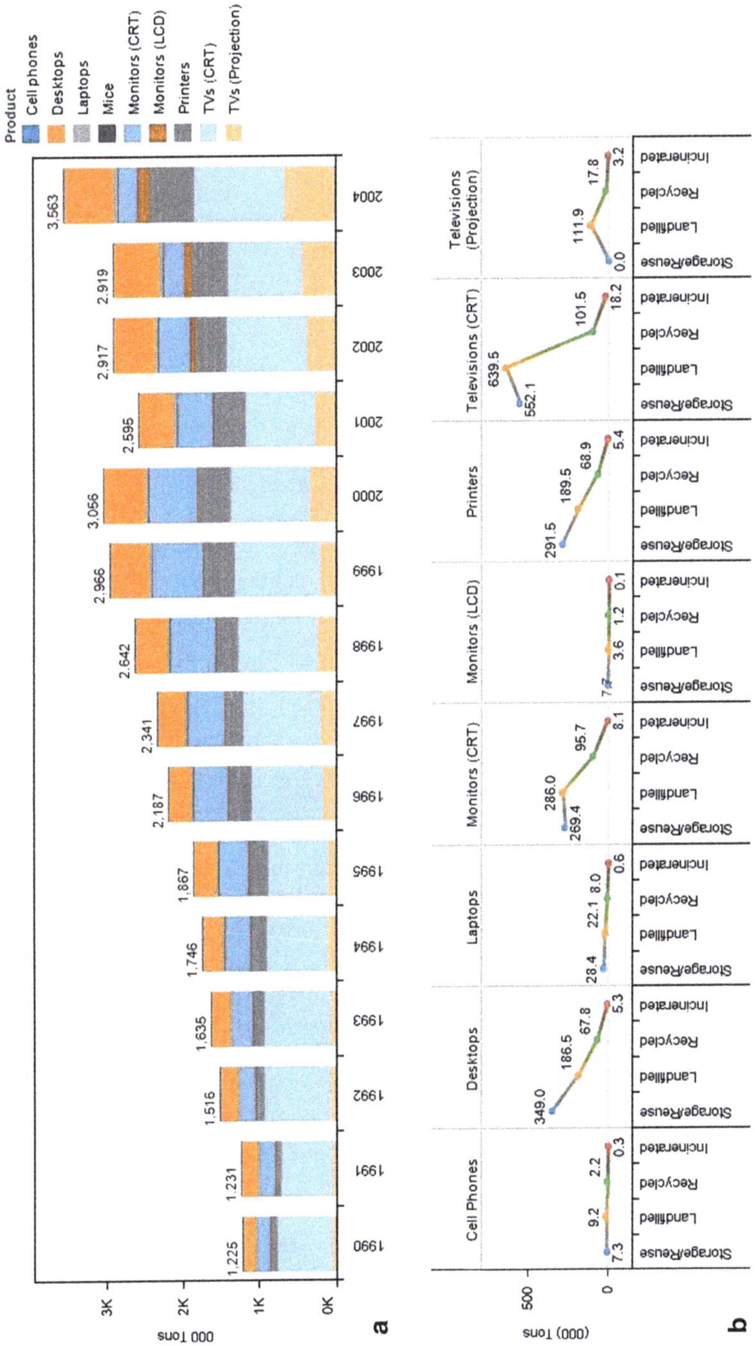

Fig. 8.2 (a) Sales volume in (000) tons. (b) End of life management of electronic devices (2005)

While e-waste represents only 2–5% of municipal waste, it is the fastest growing form of waste, not only in the US, but also globally. The United Nations Environmental Program estimates that by 2020, e-waste from old computers will have increased from two to four fold from 2007 levels in China, and five-fold in India. E-waste from discarded mobile phones will be about 7 times higher than 2007 levels in China, and 18 times higher in India. By 2020, e-waste from televisions is expected to be double from the 2007 baseline in China and India.

8.2 Why is E-Waste Such a Big Issue?

8.2.1 Material Depletion

Electrical and electronic equipment contain a wide range of materials made of metals, plastics, organic compounds, and other substances—some valuable, some toxic, some both. One concern is the high use of virgin minerals to make the EEE devices—without sufficient oversight this can lead to significant depletion of mineral resources.

Figure 8.3a, b show the typical material composition of a cell phone[5] and a computer[6]. Typical materials include base metals like copper (Cu), aluminum (Al), iron (Fe), and tin (Sn); special metals such as cobalt (Co), indium (In) and antimony (Sb); and precious metals such as silver (Ag), gold (Au) and palladium (Pd). According to UNEP[7], a single cell phone contains, for example, 250 mg Ag, 24 mg Au, 9 mg Pd, 9 g Cu, and 3.5 g Co for the battery. Considering 1.2 billion cell phones were sold globally in 2007, this amounts to 300 tons of Ag, 29 tons of Au, 11 tons of Pd, 11,000 tons of Cu, and 4500 tons of Co! UNEP estimates that in excess of 45 billion dollars of minerals are mined yearly for EEE equipment. This adds up to 30% of the world mine supply of Ag, 12% of Ag, to 13% of Pd and to 15% of Co, 50% of Sb, and 30% of Cu[8]. Given that only 15–20% is recovered to be recycled, this is a significant depletion in both base and precious metals just from EEE.

Figure 8.4a shows the estimates for different materials in electronic devices landfilled, recycled, or incinerated in the US in 2005. The chart also identifies, for each material, how much was recycled or landfilled (Fig. 8.4b). This chart is based off the statistics in Fig. 8.2b—composition and weight of each material for the range of devices in Fig. 8.2b is estimated; and the sum of each material across all relevant devices in 2005 is computed[9]. For example, 117,913,190 pounds of copper became

---

[5] "Cell phone composition." UNEP/GRID-Arendal Maps and Graphics Library. 2006. UNEP/GRID-Arendal.

[6] http://www.unep.org/publications/ebooks/kick-the-habit/.

[7] http://www.unep.org/PDF/PressReleases/E-Waste_publication_screen_FINALVERSION-sml.pdf.

[8] *Ibid. 8.*

[9] Data is from the US Environmental Protection Agency: http://www.epa.gov/osw/conserve/materials/ecycling/manage.htm.

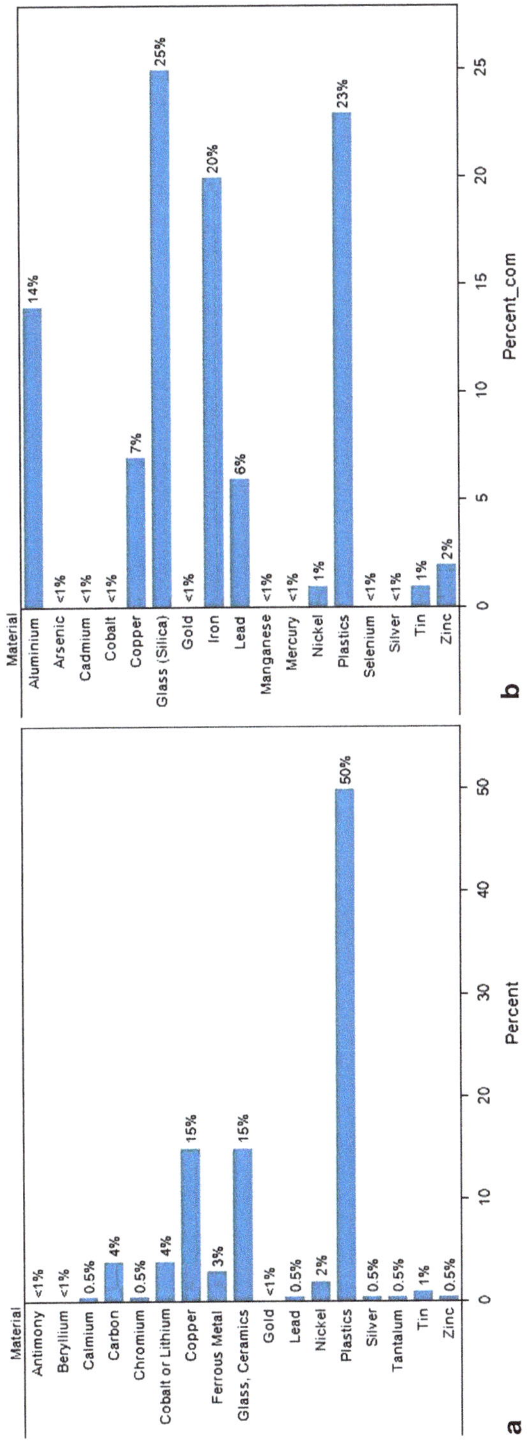

Fig. 8.3 (**a**) Typical materials in a phone. (**b**) Typical materials in a computer. (Sources: UNEP/GRID-Arendal maps and graphics library. 2006, UNEP)

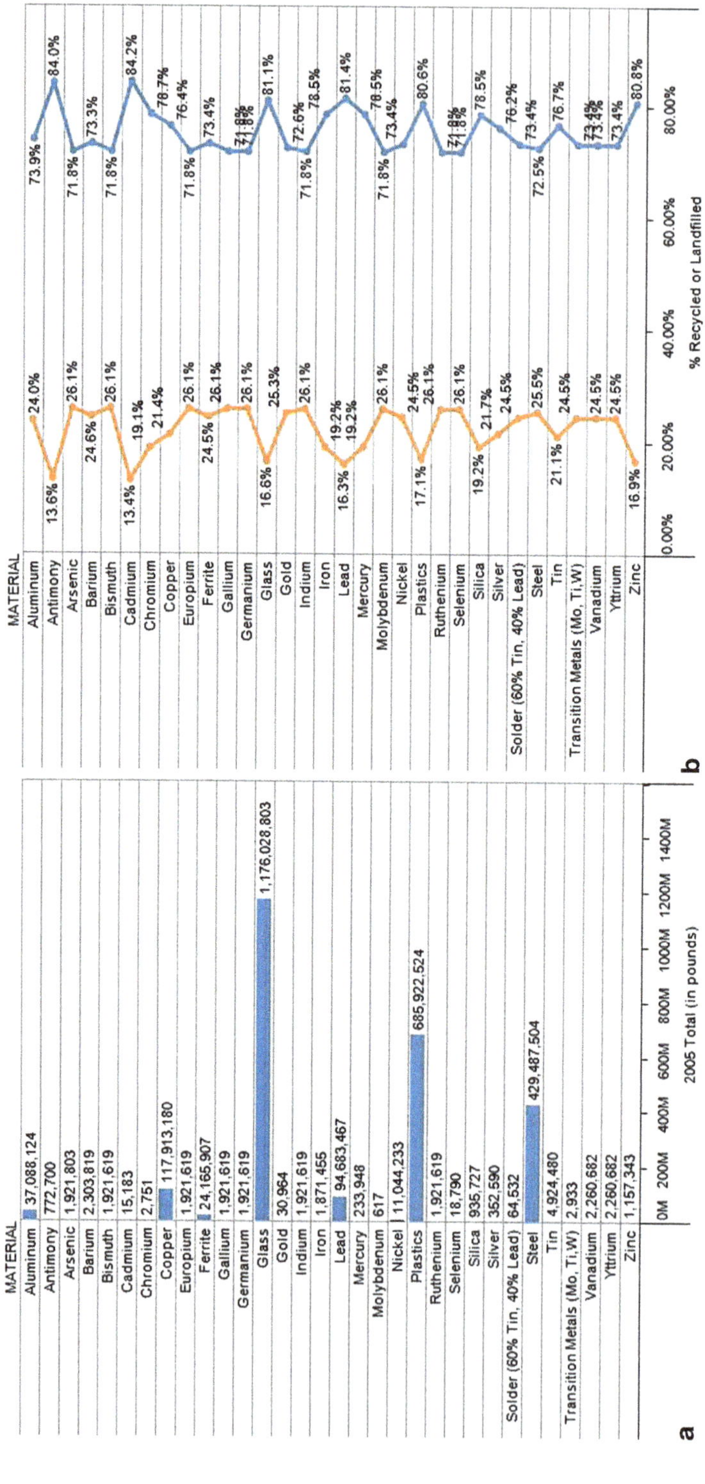

Fig. 8.4 (**a**) Total of each material at end of life. (**b**) % of each material recycled or landfilled. (Sources: US EPA)

eligible for end of life. 21.4% of that quantity was recycled and 76.4% made its way to the landfill. The remaining 2.2% was incinerated. For the sample of materials in Fig. 8.4, the recycling rate is between 13% and 27%—too low to stop significant depletion of mineral resources.

8.2.2 Energy Required for Production

EEE device production is an energy-intensive process. Metals and substances have to be mined and processed; components for the device made and shipped to the assembly sites (typically in Asia); and the completed product eventually shipped to the various markets—largest being in developed countries.

Over 60 substances are typically required for EEE devices and since only a small fraction is recovered through secondary production (recycling operations), most of the materials have to be extracted through mining operations. Figure 8.5a gives the CO_2 emissions for both primary production and recycling operations for some of the common metals in EEE devices[10]. For example, primary production of copper releases 1.25 tons of CO_2 for every ton mined. The EEE industry used 4.5 million tons of copper in 2006—it adds up to 5.625 million tons of CO_2. In contrast, recovery of copper from secondary sources releases 0.44 tons of CO_2/ton recovered.

Since the concentration of precious metals is much lower in ores, it takes a lot more energy to extract them through mining operations. For example, primary production of gold releases 16,991 tons of CO_2 per ton mined. The annual demand for gold in EEE is around 300 tons making the total CO_2 releases for gold production for EEE devices around 5.1 million tons. Figure 8.5b lists the CO_2 emissions from other precious metals.

Life cycle analysis of EEE devices indicates that production is a significant part of the carbon footprint. Figure 8.5c highlights CO_2 emissions from the manufacture of common components[11]. The manufacture of flat LCD screens, for example, has a footprint of 185 kg of CO_2. Millions of LCD monitors are sold each year; and many LCDs contain substances of concern—so in addition to the energy burden, the production and the disposal has the potential to cause significant harm to human health and environment.

Figure 8.5d shows the lifecycle of an Apple computer[12] and a typical Nokia phone[13]. The CO_2 emissions from production accounts for about two-thirds of each device's footprint. For devices with short life spans, such high production footprints are common. Washing machines and refrigerators, on the other hand, last much longer than cell phones or computers and consequently, most of their carbon footprint will be

---

[10] http://www.bir.org/assets/Documents/publications/brochures/BIR_CO2_report.pdf.

[11] http://www.unep.org/publications/ebooks/kick-the-habit/.

[12] http://www.apple.com/environment.

[13] http://www.nokia.com/environment/devices-and-services/creating-our-products/life-cycle-thinking.

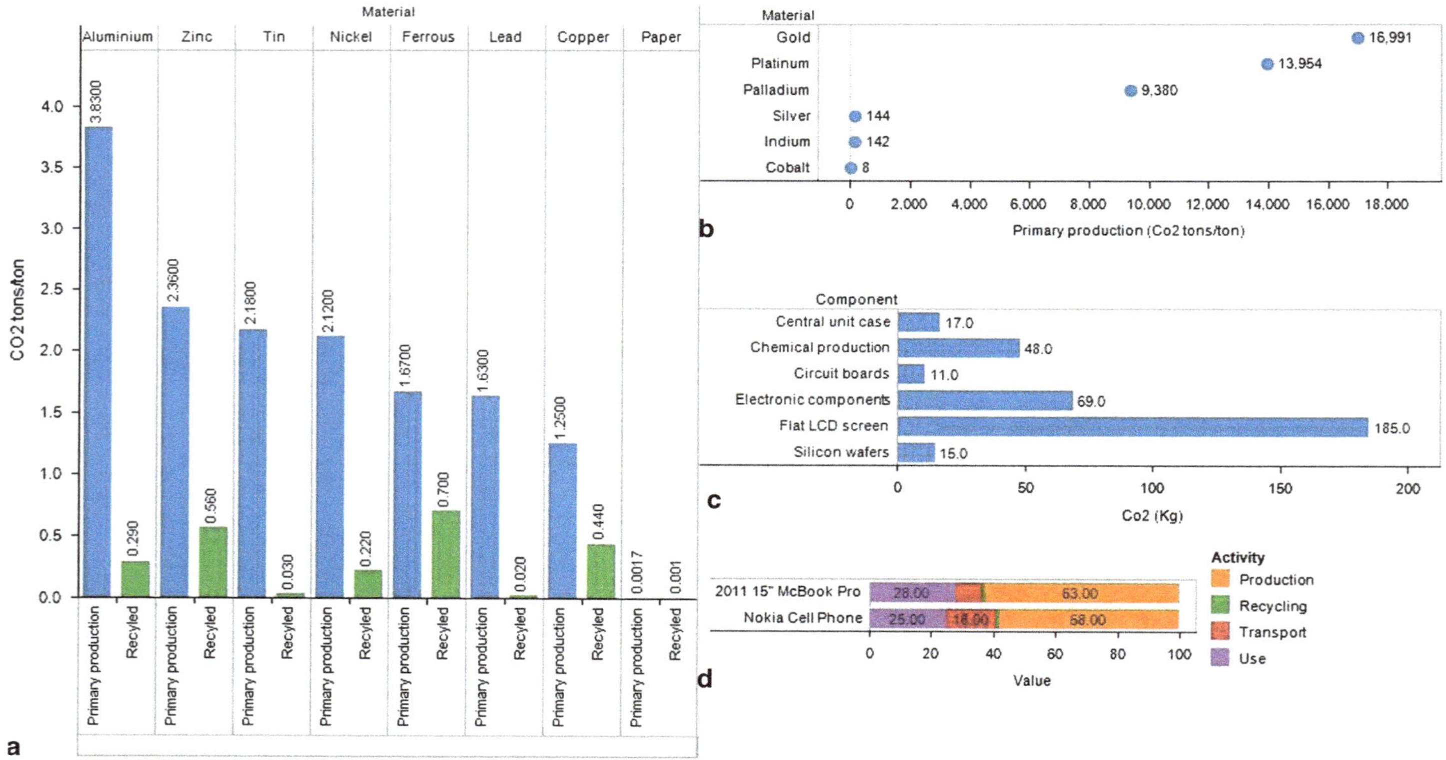

Fig. 8.5 (**a**) Energy needed for primary and secondary production. (**b**) Primary production of percious metals. (**c**) Energy needed to manufacture common computer components. (**d**) Life cycle of a computer and phone. (Data sources: Bureau of international recycling (BIR), UNEP. Apple.com. Nokia.com)

from product use. Such life cycle analysis suggest appropriate end of life management for EEE devices. Increasing a computer's lifespan by 10%, for example, can reduce its life cycle impact by 8.6% while recycling it only reduces it by.43%[14]. So incentives for reuse or extending lifespan will conserve energy and reduce e-waste. In case of refrigerators and washing machines, replacement at end of life with energy efficient appliances will save energy.

While end of life decisions are dependent on the device, the high energy needed to manufacture and use the product coupled with the low recycling rates in EEE are responsible for significant greenhouse gas emissions.

8.2.3 *Toxicity/Health Issues*

E-waste contains many toxic substances—metals such as lead, mercury, cadmium; organic compounds such as brominated flame retardants (BFR), PCB polychlorinated biphenyls (PCB), and polyvinyl chloride (PVC) that either are proven or suspected to cause damage to the environment and human health. Table 8.1 highlights some of the commonly occurring toxic substances in e-waste, their applications, and their potential for causing human harm.

Toxic substances in EEE are typically in solid and non-dispersible form, so there is little danger of human and environmental harm during the course of their normal use. The primary cause for concern is the improper management at the end of its life.

Found in plastic casings of personal computers, CRT monitors, and printed circuit boards, BFRs—including common forms such as Polybrominated diphenyl ethers (PBDEs) and Tetrabromobisphenol-A (*TBBP-A*)—can persist in the environment and bio-accumulate in living organisms, causing liver and thyroid toxicity. PCBs are organic compounds used in EEE in electric transformers and capacitors, and as additives in sealants and plastics. PCBs pose risks to the nervous, reproductive and immune systems. Both BFR and PCBs can be released into the environment when computer parts are shredded or heated. PVC is used to insulate cables and wires and release toxic dioxins when incinerated in the open.

Cathode ray tube televisions and computer monitors have lead built into the glass to stop radiation leakage. If improperly disposed, lead in CRTs can leach into the ground and eventually into the water table. Lead is also used in battery production, metal products such as solder and pipes, and devices to shield X-rays. If ingested or inhaled it can harm the nervous system, kidneys, and reproductive system.

Small amounts cadmium is present in electronic contacts and switches. Highly toxic, cadmium persists in the environment and bio-accumulates in living organisms. Printed circuit boards with CPU units, specialty chips, resistors, and capacitors may

[14] Williams, E. D., Y. Sasaki. 2003. Energy analysis of end-of-life options for personal computers: Resell, upgrade, recycle. Paper presented at the *2003 IEEE International Symposium on Electronics and the Environment*. Boston, MA: May 19–23.

Table 8.1 Toxic substances in e-waste

Substance	Typical Uses	Suspected Human Harm
PCB (polychlorinated biphenyls)	Condensers, transformers	Risks to nervous and reproductive systems
Brominated fire retardants (BFRs)	Fire retardants for plastics (printed circuit boards, connectors, cables insulation)	Several compounds suspected to be toxic
Polyvinyl chloride (PVC)	Cable insulation	Phthalates like DEHP and dioxins are suspected toxins
Arsenic	Lead shot, Electrical circuits	Toxic, Carcinogenic
Beryllium	Magnetrons, x-ray transparent windows, heat transfer fins in vacuum tubes and glass lasers	Toxic, lung disorders
Cadmium	Batteries, plastics	Damages the lungs, causes kidney disease, and irritates the digestive tract
Hexavalent chromium	Rust-resistant coating to metals/pigment in plastics	Carcinogen
Lead	CRT screens, batteries, printed wiring boards	Can harm nervous system, kidneys, and reproductive system
Mercury	Fluorescent lamps that provide backlighting in LCDs	High levels may damage the brain, kidneys, and developing fetuses.
Nickel	Batteries	Allergic reactions

also contain beryllium, arsenic, and lithium, which could cause human health and environmental concerns if improperly managed.

Mercury, used in flat screen TVs and in certain batteries and switches is highly toxic, bio-accumulates, and may cause neurological and kidney damage.

The recycling of EEE is a complex and expensive process when done right. The presence of toxic substances calls for specialized technology and regulatory controls to recover materials for future use and the safe capture and disposal of substances of concern.

8.2.4 Global Trade

According to the Basel Action Network[15]:

> There is an ugly underbelly of economic globalization that few wish to talk about. Under the guise of simply utilizing the "competitive advantage" of cheap labor markets in poorer areas of the world, a disproportionate burden of toxic waste, dangerous products, and polluting technologies are currently being exported from rich industrialized countries to poorer developing countries. In effect, rather than being helped to leap-frog over dirty development cycles directly toward clean production methods, developing countries are instead being asked to perpetuate some of the world's most toxic industries and products and are even asked to become the global dumping ground for much of the world's toxic wastes.

[15] http://www.ban.org/main/about_BAN.html.

Some argue that the export of EEE devices from Organization of Economic Co-operation and Development (OECD) to non-OECD countries is largely for reuse; still others argue that the export makes its way to responsible recyclers so EEE can be recycled cheaply and safely; but there is no denying that—despite international treaties—thousands of tons of e-waste unscrupulously makes its way to developing countries.

Most developing countries lack the capacity and appropriate infrastructure to safely manage e-waste that are either imported or locally generated. Large quantities of e-waste that are imported are improperly handled and often informally recycled in "backyard" operations "where some of the poorest laborers in the country disassemble and process them using crude, unsafe methods—like breaking open CRT monitors with hammers, 'cooking' circuit boards to remove chips, sweeping printer toner dust from cartridges and using dangerous acid baths to extract gold. Workers rarely have any protective equipment, like gloves or masks. And they are unaware of the risks of contamination to which they are exposed[16]."

In the last ten years NGOs such as the BAN[17], Greenpeace[18], Silicon Valley Toxics Coalition[19] and investigative news programs such as Frontline[20] and 60 min[21] have documented the informal recycling operations of e-waste often by children under appalling conditions in Ghana, Nigeria, India, Pakistan, and China.

The solution to the toxic trade is a combination of strong national or regional legislation on extended producer responsibility; banning toxic trade; providing producers incentives to design products that are not harmful to human health and environment; and at the end of life, promoting take back and processing that is local, safe and cost effective while creating jobs.

8.3 How Can E-Waste be Addressed? What are Countries Doing?

8.3.1 Basel Convention

The Basel Convention on the Control of Trans-boundary Movements of Hazardous Wastes and Their Disposal, commonly referred to as the "Basel Convention," seeks to protect human health and the environment from the dangers posed by hazardous wastes, including e-waste. The convention uses a three-step strategy to achieve its objectives. The first step is to promote clean production practices that minimize

---

[16] http://www.electronicstakeback.com/wp-content/uploads/Q_and_A_on_Exporting_Issues.

[17] http://www.ban.org/.

[18] http://www.greenpeace.org/usa/en/news-and-blogs/news/poisoning-the-poor/.

[19] http://svtc.org/blog/e-waste/the-plight-of-electronics-refurbishers-in-india/.

[20] http://www.pbs.org/frontlineworld/stories/ghana804/video/video_index.html.

[21] http://www.cbsnews.com/stories/2008/11/06/60 min/main4579229.shtml.

the generation of hazardous wastes. The United Nations Environment Program is working closely with businesses to identify and disseminate "best practices" as part of its efforts to promote this goal of the Basel Convention. Second, the wastes that are generated are best managed locally. This warrants strong commitment from municipalities, governments, and businesses to set up the required infrastructure to safely manage wastes. The convention has developed multiple technical guidelines to aid in this process. Finally, perhaps the most important aspect of the Basel Convention is the control of hazardous waste across international boundaries. Specifically, the "Basel Ban" amendment bans the transport of hazardous waste from the developed countries to the developing or under-developed nations.

8.3.2 Spot Light on EU's WEEE/RoHS/REACH Legislation

EU's landmark legislation on the Restriction of Hazardous Substances (RoHS, Directive 2002/95/EC[22]) restricts the use of specific hazardous materials found in electrical and electronic equipment. The banned substances include lead, mercury, cadmium, hexavalent chromium, polybrominated biphenyls (PBB) and polybrominated diphenyl ethers (PBDE). The WEEE Directive (Directive 2002/96/EC) promotes the collection and recycling of EEE equipment. The directive is based on the producer responsibility approach—it makes the producer or distributor responsible for managing the product at the end of its useful life. It provides for the creation of take back policies where consumers return their used e-waste free of charge. The Directive also mandates recycling rates on classes of WEEE. Both the RoHS and the WEEE Directives have been in force since February 2003.

Another legislation that is having a significant impact on EEE devices (and eventually e-waste) is EU's "REACH" legislation (EC 1907/2006). It deals with the Registration, Evaluation, Authorization and Restriction of Chemical substances. The law entered into force on 1 June 2007. It requires firms placing components or products in EU member countries to register all substances in their devices in a central database run by the European Chemicals Agency (ECHA). The Agency manages the database, co-ordinates the evaluation of suspicious chemicals and maintains a public database in which consumers and other interested parties can find hazard information of the substances in devices.

The REACH legislation also calls for the progressive substitution of the so-called "substances of very high concern" when suitable alternatives have been identified.

The EU is currently in the process of recasting the WEEE and RoHS with the intent of improving the recycling rates. The new proposal sets mandatory collection targets equal to 65% of the average weight of EEE introduced to the market over the two previous years in each Member State.

The EU's WEEE and RoHS Directives serve as the baseline for similar laws around the world (see Table 8.2). Japan, Canada, Australia, China, and India all have

[22] http://eurlex.europa.eu/LexUriServ/LexUriServ.do?uri=CELEX:32002L0095:EN:HTML.

Table 8.2 Summary of laws impacting e-waste

Country/ Region	Laws or Directives [signed into law]*	Key Idea
EU	WEEE Directive (2002/96/EC) [2003]	Extended producer responsibility Mandated recovery rates for specified groups of products
	RoHS (Directive 2002/95/EC) [2003]	Restricts use of lead, mercury, cadmium, hexavalent chromium (CrVI), polybrominated biphenyls (PBB) and polybrominated diphenyl ethers (PBDE)
	Registration, evaluation, authorisation and restriction of chemicals (REACH) (EC 1907/2006) [2007]	Restricts the production and use of certain chemicals
China	The china restriction of hazardous substances [2006]	Legal threshold established for lead, cadmium, mercury, hexavalent chromium, polybrominated biphenyls (PBBs), and polybrominated diphenly ethers (PBDs)
	The management regulation on the recycling and treatment of disposed appliances and electronics products [2009]	Extended producer responsibility
Japan	Law for the promotion of effective utilization of resources [1991, 2000]	Schemes similar to EU WEEE
	Home appliances recycling law [1998]	Governs recycling of home appliances
	Act on confirmation, etc. of release amounts of specific chemical substances in the environment and promotion of improvements to the management thereof [2007]	Voluntary improvement of the management of chemical substances by business operators
India	E-waste (management and handling) Rule, 2011 [2011]	Extended producer responsibility

laws that either ban hazardous substances or use the extended producer responsibility approach to promote recycling and reuse of electronic equipment.

There is currently no Federal mandate to recycle e-waste in the USA. However, twenty-five states have passed legislation on restriction of hazardous substances and recycling of EEE devices. The laws vary widely on what is covered and how it is implemented, but they typically use the Producer Responsibility approach, where the parties responsible for putting the product into the market are also responsible (partly or completely) for managing it at the end of its life. California uses the Advance Recycling Fee, where the customer pays a fee at the point of sale. These payments are used to recover and recycle e-waste by qualified recyclers. It is estimated that about 65% of the population of the US is now covered by a state e-waste recycling law.[23]

[23] http://www.electronicstakeback.com/promote-good-laws/state-legislation/.

Table 8.3 Combating e-waste

Category	Strategy
Raw material use	Use fewer inputs Use less of each input Use recycled content Use recyclable content
Elimination of Hazardous Substances	Eliminate and substitute substances of concern (RoHS/REACH compliance) Use precautionary principle Work with suppliers to establish timeline for elimination of substance of concern
Design for recovery	Design for disassembly Modular construction
Manufacturing	Use renewable energy Reduce waste
Product use	Energy efficiency Longer lasting products
Take back	Work with retailers, federal and state agencies to recover product Educate customers Have convenient "drop off" points Lobby for producer-responsibility laws

Rep. Gene Green (D-TX) and Rep. Mike Thompson(D-CA) introduced the bill SR 2284, the "Responsible Electronics Recycling Act of 2011" in Congress to make it illegal for US to send toxic e-waste to developing nations. The bill has bi-partisan support and is also backed by many firms including HP, Apple, Best Buy, and Samsung.

On 15th November 2010, President Obama issued a presidential proclamation declaring 15th November as "America Recycles Day"—to increase awareness on the issues of e-waste and to promote recycling of EEE. In the proclamation, he also announced that he was creating an Interagency Task Force within the federal government "to prepare a national strategy for responsible electronics stewardship, including improvements to Federal procedures for managing electronic products[24]."

8.4 What are Firms and Their Supply Chains Doing to Reduce E-Waste?

Firms are focusing primarily on three areas to tackle e-waste: (1) Design choices, (2) Manufacturing efficiency, and (3) Effective take-back programs. Table 8.3 gives a summary of the major initiatives.

[24] http://www.epa.gov/epawaste/conserve/rrr/2010recycles.prc.rel.pdf.

8.4.1 *Raw Material Use*

Green design or Design for the Environment (DfE) are terms used to describe design choices that reduce the impact on the environment. The primary idea is to incorporate the impact from the product's life cycle when designing a product. This would include all stages—from extraction of raw materials, manufacture of its components, product use, and eventually its take back at end of life. With burgeoning laws on extended producer responsibility around the world, designing for efficient use and recovery is becoming a necessity for many EEE firms.

When it comes to raw material use, the thrust is to use fewer materials ("dematerialization"), use less of each input, use recycled content and finally use material that can be recycled. For example, the Dell's "Studio Hybrid" desktop is about 80% smaller than standard desktops. The Dell "Optiplex 960" desktop contains 10% post-consumer recycled content plastics and many of Dell's displays contain 25% post-consumer recycled content plastic. HP prints manuals for their LaserJet printers on elementary chlorine free bleached virgin or recycled paper or puts it on disk to reduce paper use. Apple uses aluminum cases and arsenic-free glass for many of their devices for ease of recycling at end of life.

8.4.2 *Elimination of Substances of Concern*

Another major impetus for designing for the environment involves the elimination or the phase out of hazardous substances from electrical and electronic devices. EEE firms are working with suppliers first to identify chemicals and substances in components and devices, evaluate them, and substitute any substances of concern with safer alternatives.

Electronic supply chains are complex—material is moved and processed in several echelons before it reaches the manufacturer. So identifying and disclosing product chemistries of substances in the product lifecycle warrants collaboration between multiple tiers of suppliers in the supply chain.

While some substances are scientifically proven to be harmful to human health and environment (for example those that are identified in RoHS legislation), there are others that under reasonable scientific grounds are suspected to cause significant harm, but lack sufficient scientific evidence of its risks. In such cases, the "precautionary principle" is often used as a foundation—action should be taken to eliminate or substitute such substances even if the risks lack full scientific certainty.

Many substances of concern such as PVC and BFRs have well-established supply chains and suppliers will have to retool and/or develop newer materials and technologies to replace them. It is typical of environmentally progressive firms to establish a timeline to implement the elimination or reduction of such substances as newer supply chains are structured and innovative but safe alternatives are developed.

Manufacturers like Nokia, Dell, Apple, Sony Ericsson, IBM, Intel, Motorola, Panasonic, Phillips, and Sony all have substance management programs, supplier responsibility initiatives, and typically provide a timeline on when substances of concern will be substituted by safer alternatives.

For example, Nokia maintains a "Nokia Substance List"—a continually updated list of substances that they "have banned, restricted, or targeted for reduction with the aim of phasing out their use" in their devices. They work closely with their suppliers and require full declaration of the substances used in their devices.[25] In addition, they explore opportunities for using new, more environmentally friendly materials, such as bio plastics or recycled metals and plastics. All of its new models have been free of PVC since the end of 2005; and all new models of mobile phones and accessories launched in 2010 were on track, in line with their goals, to be free of brominated compounds, chlorinated flame-retardants and antimony trioxide.

Sony Ericsson products are free from PVC and BFRs, with the exception of a few components that are still being phased out. Sony Ericsson has banned antimony, beryllium and phthalates from new models launched since January 2008. Motorola is phasing out BFRs by substituting it with a halogen-free laminate that is cost effective, while meeting fire safety standards. Toshiba has replaced BFR-containing plastic casings in electronic parts with inherently flame-resistant polyphenylene sulfide.

There are several NGOs working in conjunction with businesses to monitor and phase out substances of concern, The International Chemical Secretariat (ChemSec), for example, maintains a list of substances of high concern (the SIN (Substitute it Now!) list) based on the EUs REACH criteria. Firms can use it to evaluate their substance management programs and share knowledge on safer substitutes. Several industry-wide platforms are also helping in supply chain collaboration. Phillips, for example, uses BOMCheck, a web-based platform where suppliers from multiple tiers can declare substances in their components. Phillips then cumulates all declarations so they know the concentrations of various substances in their products. They can use it for compliance (REACH declarations) and monitoring their own substance management programs.

8.4.3 Manufacture and Use

Firms are addressing the vast amounts of energy used to manufacture EEE devices; and in the eventual use of these devices in many ways. First, EEE companies are investing in supply chain efficiency and carbon foot print reductions. This entails energy efficiency and investing in renewable sources of energy. Second, efforts are also underway to manufacture devices with significant recycled content, reducing the amount of energy required to process it. Third, the products are more energy efficient and last longer, reducing the energy burden to use and replace it.

[25] http://www.nokia.com/environment/strategy-and-reports/environmental-strategy.

Dell, Apple, HP, Phillips, Motorola, have all, for example, made significant progress in mapping the carbon footprints of many of their products through its life cycle[26] so that they can focus on abatement schemes.

23% of Motorola's energy, for example, comes from renewable sources; Apple's facilities in Cork, Ireland; Austin, Texas; and Sacramento, California, converted to using 100% renewable energy[27]. HP and Google have made investments in efficient data centers and employee travel programs significantly reducing their energy consumption.

Devices have also become more energy efficient. Energy efficient components and software routines to optimize energy usage have also contributed to energy savings during the product use stage of the life cycle. Second, EEE firms are also designing products to last longer and with enough flexibility to upgrade without replacement. Advances in battery technology are making batteries last through more charge cycles; reliability of standardized components make the device last longer while using less energy.

There are also several labeling programs that identify energy efficient devices. For example, one of the goals of the US EPA's ENERGY STAR program is to reduce greenhouse gases by identifying and promoting energy-efficient devices. The Electronic Product Environmental Assessment Tool (EPEAT) is a registry of electronic devices managed by the Green Electronics Council (GEC). The GEC evaluates devices on multiple sustainability criteria that include design, production, energy use and recycling.

Finally, design is also addressing disassembly—at the end of product life, the product can be taken apart easily and more of it recycled using less energy.

For example, HP computers use common fasteners and snap-in features and avoid the use of glues and welds where feasible. This makes it easier to manufacture and disassemble the products at the end of life. HP workstations and "DC series" desktop products have a tool-less chassis for easy upgrade and recycling at end of life.

8.4.4 *Take-Back*

Take back of a product—its retrieval at the end of life—is an essential element of "closing the loop" in EEE supply chains. Companies take back products to recover substances and components or for remanufacturing or establish themselves as a market for secondary materials collected by others. In many parts of the world with extended producer responsibility laws, take-back is often mandated, so addition to economic, social, and environmental reasons for take-back programs, firms do it to comply with local laws.

---

[26] See the Chapter on "Carbon Footprinting: A Supply Chain Approach" for more details on how to reduce carbon footprints.

[27] http://www.apple.com/environment/progress/.

There is no single model for the effective take back of products. Members of the supply chain—manufacturers, retailers, consumers, NGOs, non-profits, and local and national governments share responsibility for take-back.

8.4.4.1 Customer Education

First, customers and users should be encouraged to recycle EEE devices at the end of life. In a survey of cell phone users in New York, for example, 60% said that they simply stored cell phone at the end of life. 23% donated it to charity or to family member or traded it in with their carrier. 7% trashed it, and 4% sold their phones[28]. The proportion of users who recycled it was less than 9%. Such low recycling rates are common among a broad spectrum of EEE devices. Given that large portions of such devices are recyclable (for example, 80% of the phone can be recycled), it is incumbent on local governments, retailers, and manufacturers to run campaigns to encourage recycling. Second, customers and users need to be informed on how devices can be recycled; and have access to a waste management infrastructure that is convenient to use. In many countries and municipalities, it is illegal to trash EEE devices. Local governments in partnership with businesses, qualified waste management companies, NGOs, or non-profit organizations that promote E-cycling, often have "recycling drives" to recover EEE devices and such collection days are widely advertised to the public. Retailers and manufacturers are now proving detailed information on products can be recycled or dropped off in conveniently located locations.

8.4.4.2 Retail and Manufacturer Programs

Many retailers—in partnerships with governmental agencies, NGOs, and manufacturers—offer take back programs of most EEE devices. In the US, for example, Best Buy, Staples, Radio Shack, all have programs where customers can drop off their unwanted devices or in many cases, get credit towards a purchase.

Manufacturers offer multiple channels for product take back. Many take back devices from customers directly, often free of charge, at selected drop off points or via the mail. Nokia, for example, offers take-back in 85 countries, both at their retail stores and via more than 4,500 service centers[29]. Others partner with authorized recyclers or waste management companies to collect EEE devices. For example, in addition to direct take back, LG has partnered with Waste Management (WM) to help customers recycle its EEE devices in an environmentally safe way. Customers can bring unwanted LG products to any participating Waste Management "eCycling" drop-off center and recycle them free of charge[30]. Another common model is for

---

[28] http://www.nokia.com/corporate-responsibility/cr-report-2007/environment/products-and-services/take-back-and-recycling.

[29] *Ibid. 27.*

[30] http://www.lg.com/global/sustainability/environment/take-back-recycling.jsp.

manufactures to form a "collective" to recycle devices and to comply with local laws. Panasonic, Toshiba, and Sharp founded the Electronic Manufacturers Recycling Management Company (MRM), whose objective is to "manage collection and recycling programs in the United States for electronics manufacturers and others, including state and local governments[31]." MRM is setting up a national infrastructure to collect electronic devices. So far 25 manufacturers and retailers (including Mitsubishi and Vizio) participate in MRM programs—and according to MRM 90 million pounds of electronics have been recycled since 2007[32].

Another example is the Rechargeable Battery Recycling Corporation (RBRC) established in 1995 by the members of the rechargeable nickel-cadmium battery industry. With 28,000 participating retailers, the RBRC launched a nationwide industry-funded recovery system to take back and recycle nickel-cadmium batteries.

8.4.4.3 National and International Initiatives

Many local, national, and international organizations (governments, NGOs, non-profits, etc) are also partnering with businesses and other interested parties to help promote recycling of e-waste.

For example, the US EPA has partnered with leading consumer electronics manufacturers, retailers, and mobile service providers[33] to create the "Plug into eCycling" program that "fosters and promotes opportunities for individuals to donate or recycle their electronics[34]."

The goals of the partnership are to promote recycling and reuse of electronics, conserve valuable resources, and to recycle recovered devices in a responsible way. According to the EPA, in 2008, Plug-In eCycling partners collected and recycled 68 million pounds of used consumer electronics in the US.

The Solve the E-Waste Problem (StEP) initiative is a UN supported global public-private initiative that includes charter members such as Phillips, Hewlett-Packard, Microsoft, Dell, Ericsson, and Cisco Systems, in addition to governments, NGOs, academic institutions, and recycling/refurbishing companies. Their declared aim is to "plan, initiate and facilitate the sustainable reduction and handling of e-waste at political, social, economic and eco-logical levels[35]." They serve as clearinghouse of ideas on policy, and on the reuse, recycle, and redesign of EEE devices.

---

[31] http://www.sharpusa.com/AboutSharp/SharpAndTheEnvironment/Recycling.aspx.

[32] http://www.mrmrecycling.com/index.htm.

[33] Some prominent partners include Best Buy, Dell, HP, Nokia, Panasonic, Samsung, Sony, AT & T, and Verizon.

[34] http://www.epa.gov/epawaste/partnerships/plugin/index.htm.

[35] http://www.step-initiative.org/.

8.4.4.4 Responsible Recycling

Responsible and energy efficient recycling to recover materials is a critical element after take-back. Recycling to recover material from EEE is a labor intensive, technically complex, and without controls, a dangerous process. The recycling supply chain for e-waste consists of three main steps: (1) take-back and collection of EEE (2) sorting, dismantling and pre-processing and (3) refining of recovered material and disposal of waste including toxic waste[36].

Sorting and dismantling is largely a manual process and can be done at local levels with proper training. Recovery of materials, especially precious metals that are present in lower concentrations, is technically complex, and requires high levels of investment in technology and process management to recover them. Also, controls need to be in place to capture hazardous waste and dispose them in a safe manner.

Consequently, it is important for both customers and organizations to use certified recyclers who recycle EEE devices safely and efficiently. Unscrupulous recyclers either improperly treat e-waste or simply ship EEE devices to developing countries under the guise of "reuse."

By verifying that an electronics recycler adheres to environmental, safety, and public health controls, accredited certification programs enable customers to make informed decisions and increased confidence that their end-of-life electronic equipment will be dealt with in an environmentally responsible manner. The e-Stewards certification program is an initiative launched by the Basel Action Network and it identifies "recyclers that adhere to the highest standard of environmental responsibility and worker protection[37]." Samsung, for example, is an e-Stewards "certified" company. Samsung's certification "confirms that e-waste collected within the Samsung Recycling Direct program will not be exported to developing countries, incinerated, or buried in municipal landfills.[38]" The e-Stewards Certification requires recyclers to protect consumer data throughout the recycling process—so Samsung's processes has controls to make sure that data on their devices are secure and is appropriately destroyed[39].

Developed by the EPA, States, manufacturers, recyclers, trade groups, and NGOs, "Responsible Recycling" (R2) practices is another standard adopted by the electronics recycling industry to certify recyclers. The R2 Practices are a set of voluntary best practices for the electronics recycling industry that reduce the environmental impact, improve worker health, and improve safety performance of electronics recyclers.

---

[36] *Ibid. 7.*

[37] http://e-stewards.org/certification-overview/.

[38] http://www.samsung.com/us/article/an-e-stewards-enterprise-certified-environmental-accountability.

[39] A recent investigative report by PBS' Frontline program (http://www.pbs.org/frontlineworld/stories/ghana804) showed that the export of e-waste to developing countries has given rise to data fraud—criminals getting access to customer data and using them inappropriately.

8.5 Conclusions

The solution to e-waste requires a concerted effort from all parties in the electronics supply chain. Customers of EEE, who drive the market for such devices, should demand greener products from manufacturers. Many NGOs such as Greenpeace and Electronics Take-Back Coalition rank manufacturers and devices on their environmental performance. Third party certifications such as EPEAT and ENERGY STAR evaluate devices on energy efficiency should advice product purchase decisions. The adage "Reduce, Reuse, Recycle" encourages using devices longer, promotes reusing, and at end of life, recycling it via certified recyclers.

Businesses need to redesign products with low environmental impact, streamline supply chain to reduce their footprint, and at end of life take-back, recover components and substances closing the loop in the supply chain.

Local and national governments should pass and enforce laws on extended producer responsibility, partner with businesses to implement e-waste reduction programs, and educate its citizens on how to manage e-waste at the end of life.

NGOs, non-profit organizations, international agencies also have an important role to play—informing the public, being a watch dog and a "whistle blower," establishing best practices, and lobbing for better legislation.

All parties—customers, businesses, governments, and NGOs and non-profits have made significant progress over the last decade in reducing the e-waste problem. With better laws, progressive corporate strategies, informed and demanding customers, we are confident that the volume of e-waste will significantly be reduced in the coming years.

Chapter 9
Product Take-Back Legislation and Its Impact on Recycling and Remanufacturing Industries

Gökçe Esenduran, Eda Kemahlıoğlu-Ziya and Jayashankar M. Swaminathan

Abstract Take-back legislation holds producers financially responsible for handling and treating their products at end of life. A growing number of countries around the world have enacted such legislation, especially for electrical and electronic products. Clearly, such legislation impacts the strategic and operational decisions of companies operating in the affected industries and the operations management literature recently started to analyze these problems. In this chapter, we provide an overview of existing take-back legislation and the papers that have studied various research questions associated with them. Our focus is particularly on papers that have studied the impact of these regulations on the recycling and remanufacturing industries.

9.1 Overview of Product Take-Back Legislation

In this era of information technology, customers enjoy prosperity, convenience, and the low prices of various electric and electronic products. As the demand for these products increases, the amount of such products being discarded by customers has been increasing, too. Electronic waste (e-waste), such as televisions, computers and cell phones, is the fastest growing municipal waste stream in the US. In 2006, total volume of waste increased by only 1.2%, but the total volume of e-waste increased by 8.6% (http://www.electronicstakeback.com/). US Environmental Protection Agency (EPA) estimates that 400 million units of e-waste are discarded each year and the figure is going up steadily. It is unfortunate that only 11.4% of the disposed e-waste is being treated properly. The massive amount of untreated electronic waste leaves behind lead, cadmium, mercury and other hazardous substances, and poses a threat for

G. Esenduran (✉)
Fisher College of Business, The Ohio State University, 2100 Neil Avenue,
Columbus 43210, OH, USA
e-mail: esenduran.1@osu.edu

E. K.-Ziya · J. M. Swaminathan
Kenan-Flagler Business School, University of North Carolina at Chapel Hill,
CB Number 3490, Chapel Hill 27599, NC, USA
e-mail: Eda_KemahliogluZiya@unc.edu

J. M. Swaminathan
e-mail: ms j@unc.edu

T. Boone et al. (eds.), *Sustainable Supply Chains*,
International Series in Operations Research & Management Science 174,
DOI 10.1007/978-1-4419-6105-1_9,

human health and the environment. As a remedy to this issue, legislators have been imposing product take-back legislation in many countries. These take-back laws hold manufacturers responsible for the collection and proper disposal of their products discarded by customers. Examples include the Waste Electrical and Electronic Equipment (WEEE) Directive in Europe, Japanese Specified Home Appliances Recycling Law (SHARL), Japanese PC Recycling System, and state specific e-waste take-back programs in twenty four states in the US (http://www.electronicstakeback.com).

Although these regulations receive enthusiasm from environmentalists in general, there is debate on implementation details and it is not clear exactly what type of legislation will achieve the ultimate goals of reducing the amount of waste by increasing product take-back and recovery levels and creating incentives for environmentally-friendly product designs. It is clear that the legislation's requirements affect companies' responses, which in turn, determine the success of the legislation in terms of achieving these main objectives.

One of the main differences in implementation is whether the legislation imposes individual producer responsibility (IPR) or collective producer responsibility (CPR). Under IPR, each producer is responsible for the discarded products of her own brand, whereas under CPR manufacturers of similar products collectively fulfill the requirements of legislation and pay for the total volume of collected products regardless of brand. In a take-back program implementing CPR, the total cost of compliance is generally divided among the members of the program with respect to their market shares. From the view point of producers affected by legislation, there are pros and cons of each type of producer responsibility. For example, IPR gives companies greater control over end-of-life (EOL) product management. Under IPR, companies pay for the collection and treatment of their own brand-name products only. Thus they are in a better position to recoup the benefits of their environmentally-friendlier product design. For example, companies might choose to make costly design changes to facilitate EOL product treatment and to benefit from savings in recycling/treatment costs under IPR. In short, the rationale for pursuing IPR is that it provides incentives for producers to incorporate the environmental concerns into their product designs and to make design changes that would facilitate the EOL treatment. For the same reason, recently many researchers (e.g., Dempsey et al. 2010; Atasu et al. 2009) as well as some initiatives (e.g., "IPR works" set up by a group of companies and NGOs) passionately urge/support the implementation of IPR principle. On the other hand, CPR might be preferred over IPR by producers for a couple of different reasons:

i. In some countries (e.g., the WEEE Directive in EU) CPR is allowed but only for historical waste (i.e., products put on the market before August 2005). However when it comes to implementation, many countries found it easier to allow for/impose collective systems for both historical and non-historical products. It is argued that once the historical waste is cleaned up (approximately around 2020) it will be much easier to implement IPR (Clean Production Action 2007).
ii. It is known that recycling/treatment activities turn out to be economically infeasible if the total volume of products recycled/treated is low (Tojo 2003b) which might be the case under IPR. Hence, collective systems (where collection and

treatment activities are handled collectively) benefit from economies of scale in costs of compliance (Van Rossem et al. 2006). However, one should keep in mind that IPR can still be implemented in a collective system if the actual cost of compliance is calculated for each member firm (e.g., as in Japanese SHARL) (Van Rossem et al. 2006; Dempsey et al. 2010). In the following, we discuss this issue in more detail.

There has also been some confusion about which existing compliance schemes can be classified as IPR. We note that it is not required for producers to have individually-owned collection/treatment systems for their own products in order to implement IPR. Similarly, being a part of a collective collection system where the products are collected and treated in a common facility regardless of brand does not necessarily imply that CPR is being implemented. The literature clearly states that IPR can be achieved through brand segregation under a collective collection/treatment system because these systems can be set up so that they encompass individual responsibility (Van Rossem et al. 2006; Van Rossem 2008; Dempsey et al. 2010). Therefore it is not necessary to have a separate collection/treatment infrastructure for each producer. To summarize, as long as the producers are responsible for and have control over their own-brand end-of-life products IPR can be implemented regardless of how the collection and treatment are performed. Japanese SHARL, Japanese PC Recycling System, Maine's Electronic Waste Recycling Law and ICT Milieu (Dutch take-back program for information and communication technology (ICT) equipments from 1999–2003) (Van Rossem 2008; Dempsey et al. 2010) are examples of legislation that incorporate the individual responsibility principle.

In order to fulfill the requirements of a legislation, depending on whether it imposes IPR or not, firms need to either set-up an individual take-back program or join a collective one. However, in the context of product take-back this is not the end of the matter. One of the other key questions is how to treat (or recover the value in) the end-of-life products, i.e., which end-of-life treatment alternative to choose. Depending on product characteristics and legislation requirements, there might be various different treatment alternatives. Although requirements of take-back legislation can be different from one country to another, a common characteristic is that they set minimum target levels on the amount of products that should be collected and treated properly (e.g., e-waste laws of Minnesota, New York State, Illinois, etc.). Most legislation refer to "recovery" as the end-of-life "treatment" of discarded products. Recovery includes burning with energy recovery, composting, and recycling, which is defined as the reprocessing of waste materials into the production process. Another end-of-life treatment alternative for discarded products is reuse. Reuse refers to the situation where the whole appliances or components are used for the same purpose for which they were conceived. Since reuse reduces the consumption of virgin materials, it is typically perceived as being environmentally-friendlier than recovery (or recycling). A particular type of reuse is remanufacturing, which brings the appliance to "as new" condition.

Concerns about various implementations of take-back legislation are not limited to the type of producer responsibility or to treatment options per se. There are also concerns about the impact of these regulations on particular industries. Before getting into these concerns, it would be helpful to understand the prevalent structures in the recycling and remanufacturing industries: Some companies manage/monitor the recycling of their products and feed recycled materials back into their manufacturing process. For example, HP has a strategic alliance with Noranda Inc., a third-party recycling company. Noranda occupies some space in HP plants and HP monitors the recycling in these plants. Similarly, Samsung has recycling programs for its EOL products (http://www.samsung.com/us/aboutsamsung/citizenship/). Samsung works with recyclers that have been confirmed by the e-Stewards program (http://e-stewards.org/) and re-introduces the recycled material into production process. However, when the OEMs are not engaged in recycling of their own products, often times independent recyclers might fill the niche. For example, Sanyo has no product take-bake or recycling program unless required by legislation. In this case, independent recyclers collect and recycle Sanyo products. Gazelle, for instance, is an independent recycling service provider (www.gazelle.com) who collects (also pays for) EOL products and recycles or resells them. The situation in the remanufacturing industry is not much different. Some companies such as Xerox, HP, and Apple have well established product take-back programs and offer both new and remanufactured versions of their products even in the absence of legislation. While some of these companies might indeed remanufacture in-house, others outsource remanufacturing activities but still manage the sales of the their brand name remanufactured products. For example, IBM remanufactures servers in its plant (http://www-03.ibm.com/financing/pdf/ca/en/igf4-a143.pdf). On the other hand, Apple contracts with PowerOn, a third-party company for recycling and remanufacturing of its EOL products. Still, remanufactured Apple products are sold through Apple's website under its brand-name. On the other hand, some companies like JVC and Philips do not have product take-back programs unless required by the legislation (www.electronicstakeback.com/). However, one can easily find refurbished products of these brands being sold by independent remanufacturers.

As mentioned earlier, product take-back legislation imposes specific collection and recycling targets. Therefore, no matter who recycles the EOL products, as long as the targets are set appropriately there is no reason for legislation to hurt the recycling industry. However, this is not the case for the remanufacturing industry. Some, indeed, argue that existing implementations of legislation may hurt the remanufacturing business (Gray and Charter 2007), which would be an undesirable outcome because remanufacturing is an environmentally-friendlier and more efficient way of capturing the value in discarded products than recycling/recovery. The reason behind the fear is that the legislation grants OEMs first access to discarded products through collection/recycling targets and thus limits the availability of those products to independent remanufacturers.

In this chapter, we first briefly go over the literature on papers that study product take-back in the absence of take-back legislation since these models form the building blocks of models used to study take-back legislation. In Sect. 9.3, we review papers that study IPR-type legislation and classify them with respect to how they model EOL management: in-house recycling, in-house remanufacturing, third-party recycling, or third-party remanufacturing. We also provide an overview of papers that model CPR-type and tax-based legislation. We conclude with a discussion of future research directions and main insights from the body of work on take-back legislation.

Other work that provides a review of the literature on environmental legislation includes Atasu and Van Wassenhove (2010, 2011). Different from these two papers, which provide a higher level review of the literature including some papers from environmental economics, we focus mostly on papers that explicitly model the product treatment step (recycling or remanufacturing) and provide details about modeling assumptions common to this literature.

9.2 Literature on Product Take-Back in the Absence of Legislation

The literature on closed-loop supply chains is rich and growing rapidly. Early work on closed-loop supply chains mainly focused on operational issues. The issues studied include inventory management (e.g., van der Laan et al. 2004; Toktay et al. 2000), production planning (e.g., Inderfurth et al. 2004), scheduling and shop floor control (e.g., Guide et al. 2005; Souza and Ketzenberg 2002), materials planning for remanufacturing (e.g., Guide and Srivastava 1997; Ferrer and Whybark 2001), and capacity expansion (e.g., Kekre et al. 2003). Dekker et al. (2004) and Atasu et al. (2008a) provide extensive reviews of this literature.

The papers that are of greater relevance to the literature on take-back legislation are those that model strategic aspects of product take-back. In the absence of legislation, firms take back EOL products only if there is economic value in the returned product. The economic value can be retrieved by recycling or remanufacturing the product. The early work in the closed-loop supply chain management literature mostly concentrated on models where the EOL treatment option for the product is remanufacturing. Studies on recycling exist mostly in the environmental studies and socio-economics literatures and Majumder and Groenevelt (2001) provides a review of this body of work.

The strategic decisions the original equipment manufacturer (OEM) needs to make when remanufacturing is available as an EOL option include whether the OEM will also sell the remanufactured product and how the OEM should compete with independent remanufacturers. Majumder and Groenevelt (2001) is one of the first papers in this stream of research. They model an industry where the remanufactured product is offered by both the OEM and an independent remanufacturer. Through numerical experiments they show that while the OEM wants to increase remanufacturer's remanufacturing cost, the remanufacturer has incentive to reduce OEM's remanufacturing cost to induce the OEM to produce more in the first period, which in turn would increase the number of available units for remanufacturing in the next period. A seminal work by Ferrer and Swaminathan (2006) extend Majumder and Groenevelt (2001) and analytically prove some central results on which the literature builds later on. For example, they show that if remanufacturing is profitable enough then the firm increases first period production and gives up some of the first period profit to increase the number of available units for remanufacturing in the subsequent periods. They also study multi-period models and show that the optimal strategies under the two-period model are mostly valid under longer planning horizons.

The work by Majumder and Groenevelt (2001) and Ferrer and Swaminathan (2006) on competition between a remanufacturer and an OEM has been extended in several directions. In Ferguson and Toktay (2006), the OEM's remanufacturing decision is endogenous to the model. In order to compete with the independent remanufacturer, the OEM either remanufactures or collects and disposes of cores in order to push the remanufacturer out of the market. Ferguson and Toktay provide the conditions under which each of the strategies is optimal for the OEM. Most of the literature studies the competition between a single OEM and a single independent remanufacturer. A notable exception is Debo et al. (2005), which studies the competition between an OEM and N independent remanufacturers. Ferrer and Swaminathan (2009), on the other hand, model the competition between new and remanufactured products but the same firm sells both. Since this body of work provides the building blocks of the papers that model take-back legislation, in Sect. 9.3.1.3, we describe the models from these papers in greater detail and explain how the models should be modified when the OEMs are faced with take-back legislation.

9.3 Literature on Product Take-Back under Legislation

9.3.1 Product Take-Back under IPR-Type Legislation

Electronic waste might contain valuable, recoverable materials such as gold, silver, aluminum, copper, ferrous metals, and plastics. For that reason dumping such EOL electronic products in landfills is akin to dumping money in the trash. That explains why some companies, such as Xerox and IBM, have product take-back programs even in the absence of take-back legislation. These companies continue to run their in-house recycling and/or remanufacturing efforts when faced with legislation. Next we discuss some models from the literature.

9.3.1.1 In-house Recycling

Some firms such as HP are engaged in recycling of their EOL products either through their in-house recycling efforts or through third-party service providers (which we group under in-house recycling since the original equipment manufacturer makes the recycling-related decisions and just contracts out the job), such that they reintroduce the recycled materials back into their production process. Although improving environmental image is one factor that affects company's take-back decisions, recycling efforts are not always completely altruistic. For products with valuable material content, such as cell phones or computers, which contain precious metals (e.g., gold[1] or platinum), the recoverable value remaining in discarded products would turn recycling of these products into a profitable business.

[1] For example, 1 metric ton of electronic scrap from obsolete computers contains more gold than 17 metric tons of gold ore (http://pubs.usgs.gov/fs/fs060–01/fs060–01).

Atasu et al. (2009) is one of the first papers that studies the implications of take-back legislation using a stylized model motivated by the WEEE Directive. Although they clearly state that their model is motivated by the WEEE Directive, they do not explicitly mention whether they model IPR or CPR-type of legislation. However, neither a consortium structure nor the allocation of total compliance cost in a collective system is modeled explicitly. Therefore we discuss the findings of Atasu et al. (2009) in the context of IPR-type legislation. They model a social planner who picks the collection target (β) as a percentage of sales and a producer who decides how many of the returned cores to manufacture (q_M), and what percentage of these items to collect and recycle (τ) at the end of their lives. The manufacturer incurs per unit production cost c_M but reaps the benefit from recycling per unit as b_R. The total cost of collecting τq_M units is $c_C \tau^2 q_M$ and thus it is convex and increasing in collection rate τ. The underlying assumption is that per-unit collection cost increases as the collection rate increases, because it gets increasingly difficult to reach consumers. The manufacturer maximizes her profit

$$\begin{aligned} &\underset{q_M,\tau\geq 0}{Max}\, \Pi_M = q_M(p_M - c_M + b_R\tau - c_C\tau^2) \\ &s.t. \quad \beta \leq \tau \leq 1 \end{aligned} \tag{9.1}$$

where the price of the product is $p_M = 1 - q_M$. Constraint (9.1) ensures that the collection rate is higher than the target imposed by legislation (i.e., β) and smaller than 1. Atasu et al. omits the case of linear collection cost because it leads to a binary solution that depends on the sign of $b_R - c_C$ (i.e., if $b_R - c_C > 0$ then $\tau^* = 1$ and otherwise $\tau^* = \beta$) with this formulation. They provide the optimal decisions of the manufacturer under collection target β in Lemma 1 (pp. 255). For the sake of brevity, below we only present their results regarding the optimal collection rates:

Lemma 1 (Atasu et al. 2009). When $\beta \leq \frac{b_R}{2c_C} \leq 1$ then $\tau^* = \frac{b_R}{2c_C}$. However, when, $\frac{b_R}{2c_C} \leq \beta \leq 1$ then $\tau^* = \beta$. Otherwise, $\tau^* = 1$.

Social planner, on the other hand, chooses the optimal collection target β that maximizes the social welfare which is the sum of manufacturer's profit, social surplus and environmental savings. Lemma 1 implies that when $\beta \leq \frac{b_R}{2c_C}$, choosing a collection target becomes unnecessary because the optimal collection rate is already higher than the target. Therefore, Atasu et al. solve the social planner's problem for the case where $\frac{b_R}{2c_C} \leq \beta$.

$$\begin{aligned} &\underset{\beta}{Max}\, W = \Pi_C + \Pi_M + \Pi_E \\ &s.t. \quad \frac{b_R}{2c_C} \leq \beta \leq 1 \end{aligned}$$

where total consumer surplus is $\Pi_C = \frac{(1-q_M^*)q_M^*}{2}$ and the environmental savings are $\Pi_E = -\varepsilon(1-\beta)q_M^*$ where ε is the environmental cost of not recycling a product. They show that when $\varepsilon = 0$, $\beta^* = \frac{b_R}{2c_C}$. Therefore, when recycling is profitable, the authors deduce that the socially optimal collection target/rate is positive even when the environmental hazard of not recycling is zero. When the recycling is profitable, unit manufacturing cost of the manufacturer decreases and thus the manufacturer's profit increases. The decrease in manufacturing cost results in lower prices and thus higher consumer surplus. Finally, the total hazard to environment is always reduced by higher collection/recycling rates. Therefore, the total welfare increases when the recycling is profitable.

For some products, such as televisions (HECC 2005; Sodhi and Reimer 2001), the benefits generated from using recycled materials may outweigh the total cost associated with recycling, but for others the net benefit is negative. Atasu et al. also solves the cost model where recycling does not bring any benefits, but it is a costly endeavor. For the cost model, in addition to the monopolistic scenario they also solve the oligopoly scenario with n players. While some findings carry over to the cost model, others do not. For example, different than the benefit model, when the environmental hazard is zero the optimal collection target is zero under the cost model. On the other hand, similar to benefit model, they find that as environmental hazard increases or collection cost (c_C) decreases, optimal recycling target increases.

9.3.1.2 In-House Remanufacturing

For the firms engaged in product take-back, remanufacturing might be a profitable EOL treatment alternative depending on the product characteristics and condition. However, there are some industry and market characteristics that might affect the attractiveness of remanufacturing. First, the amount of discarded products available for remanufacturing is bounded by the amount of new products sold by the firm in the earlier periods. Therefore, there is a natural limit on the maximum amount of remanufacturing a firm might achieve in each period. Second, the firms are typically concerned about the cannibalization of new product sales by the remanufactured versions. Therefore, it is important to understand how customers perceive remanufactured products as compared to newly manufactured products. Next we review the assumptions about the availability of remanufactured products and then we discuss how the assumption regarding customers' perception of remanufactured products lead to the generally utilized demand model in the literature.

The first step in a product take-back program is the used product return/collection and it is the only input to the remanufacturing operations. In an industry regulated with take-back legislation, the manufacturer is responsible for collection and treatment of all products discarded by her customers. Therefore, the remanufactured product supply depends on the product sales in the previous period. Ideally all products discarded at the ends of their useful lives are available for collection and treatment. Depending on the industry, customer behavior and market characteristics, there might be uncertainties in quality, amount and timing of product returns. These elements of uncertainty in used product collection have been studied in the literature. For example, the variability and/or uncertainty in used product condition is considered in Ray et al. (2005), Galbreth and Blackburn (2010), Galbreth and Blackburn (2006),

Guide et al. (2003), the uncertainty in the amount of product returns is considered in Ferguson et al. (2009), and the uncertainty in the timing of returns is considered in Souza and Ketzenberg (2002), and van der Laan et al. (1999).

In regards to how the remanufactured products are perceived by customers compared to newly manufactured products, literature uses two different assumptions, each of which could be true for specific types of products. Some papers in the literature assume that the remanufactured products are perfect substitutes for newly manufactured products. However, the more recent and prevalent assumption is that remanufactured products are perceived as lower quality or older technology version of the newly manufactured product.

The deterministic demand model we present next is widely used in the literature. The underlying assumptions are: First, for a new product, the customer's valuation, i.e., willingness-to-pay, is uniformly distributed between 0 and Q_t where Q_t is the potential market size in period t. Each consumer is defined by her valuation of the new product, $v_t \in [0, Q_t]$ in period t. When a customer with valuation v_t purchases a new product priced at p_{tM} her utility is $v_t - p_{tM}$. Second, a customer's valuation for the remanufactured product is a fraction, denoted by $\alpha \in [0, 1]$, of her valuation for the new product. Therefore, when a customer with valuation v_t purchases a remanufactured product priced at p_{tR} her utility from the remanufactured product is $\alpha v_t - p_{tR}$. Finally, a customer purchases the product that provides her the highest utility. Since each customer maximizes her utility, using participation and incentive compatibility constraints it is easy to show that the inverse demand functions for the new and remanufactured products respectively are as follows:

$$p_{tM} = Q_t - q_{tM} - \alpha q_{tR} \tag{9.2}$$

$$p_{tR} = \alpha(Q_t - q_{tM} - q_{tR}) \tag{9.3}$$

where q_{tR} is the amount of remanufactured product and q_{tM} is the amount of manufactured products in period t. For a detailed derivation of the demand functions from the consumer utility functions see Ferguson and Toktay (2006) or Ferrer and Swaminathan (2006). In the rest of this chapter, in accordance with the literature, unless otherwise stated we assume that $Q_t = Q$. Note that, when $\alpha = 1$ the customers value the remanufactured and new products the same, i.e., do not differentiate between new and remanufactured products (perfect substitutes). Therefore this demand model can be used to model either of the assumptions regarding the substitutability of new and remanufactured products. Still, it is worth emphasizing that there are not many examples of perfect substitutes in the market. Retreaded tires for commercial fleet companies and single-use cameras are some of the examples (Souza 2008; Atasu et al. 2010a).

Next we provide a common approach for modeling the monopolistic situation where the OEM is the only party remanufacturing discarded products of her own brand. Esenduran et al. (2011) studies this problem in detail and investigates the impact of take-back legislation on an OEM with in-house remanufacturing capabilities. Monopolist OEM offers both newly manufactured and remanufactured products. Note that the OEM might have in-house remanufacturing capability or she might outsource the remanufacturing activity; however in either case the OEM has full control over the fate of remanufactured products and she sells these products under her brand-name. In a two-period setting the OEM decides how many units of new products to manufacture in each period (q_{tM} for $t = 1, 2$), how many cores to collect in the second period (q_{2C}), of the collected cores, how many to remanufacture (q_{2R}). Let c_M be the unit cost of manufacturing, c_R the unit cost of remanufacturing, c_D the unit cost of disposal and c_C unit cost of collection. Then the OEM's problem is formulated as follows:

$$(P1) \qquad \underset{q_{1M}, q_{2M}, q_{2R}, q_{2C} \geq 0}{Max} \quad \Pi_{1M} = q_{1M}(Q - q_{1M}) + \phi \begin{pmatrix} q_{2R}(p_{2R} - c_R) + q_{2M}(p_{2M} - c_M) \\ -q_{2C}c_C - (q_{2C} - q_{2R})c_D \end{pmatrix} \tag{9.4}$$

$$s.t. \quad q_{2C} \leq q_{1M}$$

$$q_{2R} \leq q_{2C} \tag{9.5}$$

where p_{2M} and p_{2R} as in (2) and (3), ϕ is the discount factor. Note that in the absence of legislation there is no incentive to acquire cores unless they are being remanufactured. Therefore total amount of remanufacturing would be equal to total amount of collection unless a product take-back legislation is in force.

Next we discuss how problem (P1) is modified under IPR-type legislation. When there is legislation that imposes a collection target on the OEM as a percentage of the output of the company, it is OEM's responsibility to pay for the proper collection and treatment of at least 100β percent (β is imposed by the legislation) of the items sold in the previous period. For example, Minnesota sets β to 80%. Therefore under a regulated market OEM's problem would change to account for the requirements of legislation. There are two modifications to formulation of problem (P1). First, we introduce a constraint that ensures that the amount of collected cores is at least as much as required by legislation, i.e., $q_{2C} \geq \beta_{q_{1M}}$. Second, the OEM needs to pay for the treatment and collection of 100β percent of its second period production. Thus, we impose the additional cost of doing so, i.e., $q_{2M}\beta(c_C + c_D)$, in the objective function. Here we should point out that the above-formulation builds on the assumption that the items sold in the second period cannot be remanufactured. This is a reasonable assumption because take-back legislation in general covers electronic products that change significantly over the course of a short period of time, thus old cores cannot be manufactured after a certain period of time.

Solution to this problem yields some interesting insights (Esenduran et al. 2011). First, if the cost of manufacturing is very low, legislation on collection levels alone neither induces remanufacturing nor creates incentives to make design changes that would lead to reduced remanufacturing costs. Therefore, policymakers should be aware that for products that are cheap to manufacture, imposing collection targets alone will not be sufficient to induce remanufacturing. However, targets on collection levels alone may stimulate remanufacturing of a high percentage of the collected cores if they are correctly chosen given the product's cost characteristics and the customers' valuation of remanufactured products. In that case, legislation also creates incentives for the OEM to reduce the remanufacturing cost unless the cost of manufacturing is too low. Hence imposing a single collection target that covers a wide range of products would not be effective if the aim of the legislation is to provide incentives also for remanufacturing of these products.

9.3.1.3 Independent Remanufacturing

In the previous section, we discussed that when the OEM sells remanufactured versions of her own-products some of the new product sales might be cannibalized. On the other hand, even when the OEM chooses not to offer remanufactured versions of her products a third-party remanufacturer might (Ferguson 2009). When a third-party remanufacturer collects, remanufactures and sells the remanufactured products, this introduces competition with the newly manufactured products of the OEM. This issue has been widely studied in the literature. Ferrer and Swaminathan (2006) is one of the first papers to study this issue analytically. They characterize the Nash equilibrium between an OEM and a third-party independent remanufacturer in a two period setting. They find that if remanufacturing is profitable then the OEM increases the first period production in order to increase the amount of cores available for remanufacturing in the second period. This result holds regardless of the OEM being a monopolist or facing competition from an independent remanufacturer. Besides, they show that as the intensity of competition increases, the OEM utilizes all available cores and offers remanufactured products at lower prices. Their results also continue to hold under multi-period and infinite planning horizons. Similarly, Majumder and Groenevelt (2001) consider the pricing and remanufacturing decision of an OEM facing competition from a third-party independent remanufacturer. They numerically show that it might be to the benefit of the remanufacturer to cooperate with the OEM to reduce OEM's remanufacturing cost. In another paper, Groenevelt and Majumder (2001) also model the competition on the procurement of cores. Ferguson and Toktay (2006), on the other hand, consider pricing, collection and remanufacturing decisions of an OEM facing competition from a third-party remanufacturer. They show that an OEM may deter the entry of an independent remanufacturer by either introducing a remanufactured product or collecting cores but not remanufacturing. They call this "preemptive collection". In addition to the competition introduced by a single independent remanufacturer, the industry with multiple independent remanufacturers has also been studied in the literature. Debo et al. (2005) investigate the joint technology selection and pricing decisions for new and remanufactured products for an OEM competing with N independent remanufacturers and compare their results with monopoly situation. They find that OEM increases the sales of new products to generate the supply of remanufactured products. Furthermore, for an OEM competing with N independent remanufacturers the sufficient condition for the introduction of remanufacturable products is the same as the one for a monopolist OEM. The optimal level of remanufacturability, however, might be lower than the one offered by a monopolist OEM. Although the competition between an

OEM and a third-party remanufacturer is studied in depth in the literature, the same is not true for the competition between two OEMs with remanufacturing capability. Heese et al. (2005) is one of the few papers that explore the competition between two OEMs. They show that for the first-moving firm remanufacturing would be a profitable option. Atasu et al. (2008b), on the other hand, show that remanufacturing might become an effective marketing strategy under competition. Especially, they show that remanufacturing is the most attractive in the presence of a strong brand image competitor.

In this section, we provide an approach commonly used for modeling the competition between an OEM and an independent remanufacturer. In the literature the competition is typically modeled as a Cournot game (e.g., Atasu et al. 2009; Ferguson and Toktay 2006; Majumder and Groenevelt 2001). Although Bertrand competition has also been used occasionally (e.g., Ferrer and Swaminathan 2006), we only provide the formulation for Cournot competition, because it is more relevant for modeling the industries regulated by product take-back legislations. Recall that Cournot competition is useful in modeling capital-intensive industries where production capacity is relatively fixed, such as electronics and automotive that are affected by legislation, whereas Bertrand competition is appropriate when the firms can easily adjust their capacity and production quantity, such as information goods.

Esenduran et al. (2011) studies the effect of legislation on an OEM who faces competition from a third-party remanufacturer. They consider a scenario where the OEM has a product take-back program but is not involved in remanufacturing. For the products she takes back, she incurs collection and proper treatment cost, where treatment involves recycling or environmentally sound disposal. The independent remanufacturer, on the other hand, might collect cores from the market or buy them from the OEM. Here we assume that the OEM has first access to the cores in the market and thus she is the leader in the Cournot competition. This is a reasonable and prevalent assumption because the OEMs has usually well established distribution channels and stronger relationships with their customer. Therefore, it would be easier and cheaper for the OEMs to reach their customers and collect the used products back. Because of the same reason, like Ferguson and Toktay (2006), we also assume that $c_C^R > c_C$, where c_C is the collection cost of the OEM and c_C^R the cost of collection for the remanufacturer.

The independent remanufacturer maximizes his profit by choosing the amount of cores purchased from the OEM, q_{2S} and the amount of cores collected from the market q_{2C}^R. Since there is no incentive to acquire cores unless remanufacturing is profitable, the remanufacturer remanufactures all the cores on hand, namely the total amount remanufactured is given by $q_{2R} = q_{2S} + q_{2C}^R$. Since the legislation is imposed on the OEM, regardless of the form of take-back legislation under which the problem is analyzed the remanufacturer maximizes his profit

$$(P2) \quad \underset{q_{2C}^R, q_{2S} \geq 0}{Max} \; \Pi_{2R} = q_{2R}(p_{2R} - c_R) - q_{2S}p_{2S} - q_{2C}^R c_C^R$$

$$\begin{aligned} s.t. \quad & q_{2R} = q_{2S} + q_{2C}^R \\ & 0 \leq q_{2C}^R \leq q_{1M} - q_{2C} \end{aligned}$$

where p_{2S} is the price at which the OEM sells the cores to the remanufacturer. The remanufacturer's constraint ensures that the quantity collected by the remanufacturer cannot exceed the amount left in the market after the OEM collects. The OEM, on the other hand, maximizes her profit by choosing how much to manufacture, how much to collect and how much to charge for the cores:

$$(P3) \quad \underset{q_{2M}, q_{2C} \geq 0, p_{2S}}{Max} \quad \Pi_{2OEM} = q_{2M}(p_{2M} - c_M) - q_{2C}c_C + q_{2S}^{*}p_{2S} - \left(q_{2C} - q_{2S}^{*}\right)c_D$$

$$s.t. \quad q_{2S}^{*} \leq q_{2C} \leq q_{1M} \tag{9.6}$$

where we use the superscript (*) to denote the optimal value of the corresponding decision variable. Note that OEM is the only player offering a product in the first period, therefore she maximizes her profit $\Pi_{1OEM} = q_{1M}(Q - q_{1M} - c_M) + \phi\Pi_{2OEM}$.

When the legislation does not impose any collection target, the above formulation remains valid. However, under the form of legislation that sets an explicit collection target β as a percentage of the earlier period sales we need to modify the formulation slightly. As the OEM needs to ensure the collection and proper treatment of 100β percent of the first period sales, we introduce the constraint $\beta q_{1M} \geq q_{2C}^{R*} + q_{2C}$. As there is no nonnegativity constraint on p_{2S}, the model allows the case where the OEM pays the remanufacturer to take the cores.

Solution of this problem reveals some interesting results (Esenduran et al. 2011). First one can show that legislation still creates incentives for the OEM—even though she is not remanufacturing in house—to reduce the remanufacturing cost unless the cost of manufacturing is too low. As for the concern regarding the effect of take-back legislation on remanufacturing levels, imposing very high collection targets may indeed cause a decrease in remanufacturing levels and hence the concerns raised by environmentalist groups are not unfounded. The same problem can be solved under the multiple period scenario. Then results show that when the product's life cycle is longer than two periods and the potential market size changes over the product's life, remanufacturing is not a very profitable business. Also, legislation with collection targets fails to induce higher levels of remanufacturing or incentive for designing environmentally-friendlier products. Furthermore, results show that legislation might incentivize the OEM to push the remanufacturer out of the market through preemptive collection in an effort to sell as many new products as possible while the demand for the product lasts. Therefore in order to avoid unintended consequences, policymakers should be vigilant before imposing high collection targets in industries where third-party remanufacturers remanufacture.

Webster and Mitra (2007) also studies the competition between an independent remanufacturer and an OEM. They characterize the Nash equilibrium for price competition in the absence of take-back legislation and generate insights regarding legislation numerically by changing the values of cost parameters to mimic different types of legislation. They show that, surprisingly, if the government is responsible for collecting the discarded products then both the OEM and the remanufacturer may benefit from legislation in terms of higher profits when compared to the no take-back scenario.

9.3.1.4 Independent Recycling

Manufacturers may outsource their recycling activities to independent third-party recyclers. Especially under take-back regulations, where the OEM is responsible for proper treatment of collected items, it is common to contract with independent recyclers to have the cores recycled properly. In such cases, the OEM sends the cores to the recycler and pays a recycling fee. Her role (or involvement) in the product's life ends at this point and she does not interfere with the recycler's decisions/processes again.

Even though it is quite common for third-parties to carry out the recycling activities, few papers model the recyclers explicitly and the research questions addressed are a variation of the following: given a total demand for recycling, how should these jobs be routed across recyclers in order to minimize costs or maximize revenues. Sodhi and Reimer (2001) models the revenue maximization problem of a recycler as a nonlinear, mixed integer program. In their model, the recycler decides what products to recycle and which smelters to use. Nagurney and Toyasaki (2005) generalizes Sodhi and Reimer (2001) and models a e-cycling network framework, which consists of e-waste sources, recyclers, processors, and consumers. Walther et al. (2008) is motivated by the existing recycling networks in Germany. These networks consist of independent recyclers and a focal company who negotiates with OEMs and allocates the recycling jobs to the member recyclers. The negotiation process is modeled as a mixed integer program and a Lagrangian relaxation based heuristic is proposed.

Toyasaki et al. (2010) approach the problem at a more strategic level and analyze a supply chain where the OEMs outsource the recycling activities entirely and model a two-stage game with two manufacturers and two recyclers. They compare two different settings: one where the recyclers are independent which they call the competitive scheme versus one where a non-profit body allocates the waste products to the recyclers which they call the monopolistic scheme. Their aim is to understand the impact of consolidation and competition in the recycling industry on recycler and manufacturer profits. They conclude that the competitive scheme often performs better because it achieves lower product prices as well as higher recycler and manufacturer profits.

9.3.2 *Product Take-Back under CPR-Type Legislation*

Although there is a strong push for IPR (Tojo 2003a; IPR Works Press Release 2007; Sander et al. 2007; Dempsey et al. 2010) from legislators, non-profits and some companies like HP and Electrolux, CPR-type implementations are also prevalent due to reasons we stated in Sect. 9.1. Therefore, it is essential to understand how effective CPR-type legislation is and how well the objectives of take-back legislation (in terms of achieving higher collection rates and incentivizing environmentally-friendlier product design) will be met under CPR. For example, it is argued that collective schemes do not give incentives for designing environmentally-friendlier products as producers do not bear the true costs of their own products. The most passionate discussion in this context has been regarding the transposition of the WEEE Directive into law. For products put on market after August 2005, the WEEE Directive (Article 8.2) not only allows producers to have access to their own products but also holds them financially responsible only for their own brand-name products. Article 8.2 in the WEEE Directive states that:

> For products placed on the market later than 13 August 2005, each producer shall be responsible for financing the operations referred to in paragraph 1 relating to the waste from his own products. The producer can choose to fulfill this obligation either individually or by joining a collective scheme.

Therefore, the WEEE Directive allows and encourages IPR. However due to imprecise transposition of the Directive into law in some EU countries, CPR is allowed and the preferred implementation (e.g., in Bulgaria, Denmark, Finland, France, Greece, Portugal, Spain, etc. (Sander et al. 2007)).

Under the take-back law in Maine, producers are given two options: (i) If the total return share is more than 5%, they can start an individual or a collective scheme; (ii) otherwise they must join the collective scheme to manage the collection and treatment of their discarded electronic products. In response to the alternative compliance schemes allowed by legislation, companies may choose different ones at different locations. For example, Dell opted to be a member of the collective program in Maine while reserving the right to update its decision. However, elsewhere (e.g., in Maryland) Dell established its own individual take-back program.

Due to the complexity of the problem, the comparison of IPR and CPR-type legislation has not been explored much in the literature. One exception is Esenduran and Kemahlioglu-Ziya (2011) where the authors compare the individual and collective compliance schemes and identify the market and operating parameters that make one scheme more cost effective than the others from the view point of manufacturers. Under the individual compliance scheme, each producer sets up her take-back and recycling systems. Under the collective compliance scheme, producers form a consortium to carry out their responsibilities under law collectively. An important issue under collective compliance is how the total cost will be allocated to the consortium members. The most prevalent approach is to allocate costs in proportion to the partner firms' market shares (MS). Another approach, which we call individual financial responsibility (IFR), allocates each producer her true share of compliance cost. Compliance schemes are compared by using a stylized model of each compliance structure. Among other results, authors show that collective compliance with IFR is, in general, the most cost effective alternative for producers because it allows producers to exploit the economies of scale inherent in a collective system as well as recoup the benefits of their environmentally-friendlier products. However, collective scheme with IFR is not easy to implement in practice because it requires sorting the end-of-life products by brand and tracking them through the treatment process in order to record the true cost of treatment. For that reason, the individual scheme and collective scheme with cost allocation by MS are the two more common implementations of compliance in practice. The authors also show that which of these two gives lower compliance cost depends on the base collection rate maintained by the municipalities and non-profits as well as the market shares of partner firms. If the partner firms have large market shares then collective compliance, in general, yields lower cost for a producer. The producer not only benefits from economies of scale but also invests in increasing the treatability level, and thus decreases the treatment cost, with no fear that the others will free-ride on her investment. As for environmental benefits, they find that the collective scheme with IFR provides superior environmental outcomes than the others, but as evidence from the implementation of the WEEE Directive suggests, this scheme is difficult to implement and enforce. On the other hand, the treatability level is higher under the individual scheme and the collection rate is higher in the collective scheme with cost allocation by MS. Finally they find that high collection targets cause producers to switch to the collective scheme with cost allocation by MS and may result in lower treatability levels. Therefore, policy makers need to understand the trade-off between the collection rates achieved and the incentives producers have to increase the treatability levels of their products.

Another paper that compares individual and collective compliance schemes is Atasu and Subramanian (2009). They model two profit-maximizing manufacturers in a vertically differentiated market where one manufacturer is the high-end producer and the other the low-end. The brand positions are exogenous to the model. The paper studies how individual and collective compliance systems affect manufacturers' recyclability choices and profits. This paper is different from Esenduran and Kemahlioglu-Ziya (2011) in two aspects. In Atasu and Subramanian, the manufacturers do not set their collection rates, but pay a per-unit treatment fee for each product sold. In addition, in allocating the collectively-incurred treatment cost to the manufacturers, Atasu and Subramanian use an average treatment cost where the weights with which each manufacturer contributes to treatment cost is exogenously set. Both Esenduran and Kemahlioglu-Ziya and Atasu and Subramanian conclude that IPR-type legislation provides better incentives for highly recyclable products. However, by making product take-back levels endogenous to the model, Esenduran and Kemahlioglu-Ziya is also able to identify a tradeoff between collection and recyclability levels.

Gui et al. (2010) also study collective compliance schemes and model cost allocation by return share. First, they show that cost allocation by return share may be perceived as unfair by some members of the consortium and result in disintegration of the collective system. Then using concepts from cooperative game theory, they propose modifications to the return share cost allocation rule and show that the resulting cost allocation schemes are in the core of the game under some mild conditions. Hence, the stability of the collective system is guaranteed.

9.4 Environmental Legislation that Taxes Consumers

The main focus of this chapter has been environmental legislation that requires producers to take their EOL products back and cover the related costs according to the extended producer responsibility principle. Another type of legislation taxes consumers either when the product is purchased or at end of life. The collected tax is then used to finance take-back and recycling programs run by governments. Here, we provide a brief overview of such legislation and the related work in the operations management literature.

The best known example of tax-based legislation is the electronic waste recycling fee charged in California. The tax is imposed on products with video display devices and, depending on the size of the video display, ranges from $ 6 to $ 10 (as of 1/1/2011) (http://www.boe.ca.gov/sptaxprog/ewfaqsgen.htm).

Even though the literature on take-back legislation is growing, we are aware of only two operations management papers that have studied tax-based systems. Atasu et al. (2010b) models both tax-based legislation and legislation that enforces manufacturers to achieve mandated take-back rates. The paper compares the preferences of different stake holders, namely the manufacturers, the consumers, the social planner and the environment, and identifies the conditions under which one type of legislation is preferable to the other. The authors show that tax-based systems are more effective for products with high environmental cost while take-back legislation is preferred by manufacturers of products with lower environmental cost relative to take-back costs.

Plambeck and Wang (2009) also model both tax-based systems and legislation that imposes individual producer responsibility. They compare these two types of legislation in terms of their impact on new product introduction. They find that tax-based systems reduce the frequency of new product introduction and, hence, increase manufacturer profits. However tax-based systems do not provide incentives for design for remanufacturability. On the other hand, legislation that enforces individual producer responsibility does provide remanufacturing incentives but fails to reduce the frequency of new product introductions. Therefore, the regulators must make a trade-off between two environmental goals in choosing between these two types of legislation.

9.5 Conclusions and Future Research Directions

Fueled by the increasing number of countries enacting legislation on product take-back, the operations management literature on such legislation has been growing. A common feature of work to date has been the use of stylized analytical models and the research questions studied have been of the strategic type such as what type of legislation should regulators prefer and when, what type of legislation is most cost effective for producers, what type of legislation provides better incentives for remanufacturing and/or remanufacturability, etc. While this line of work generated invaluable insights regarding how environmental legislation affects various decisions in electronics supply chains, we believe that future research should concentrate more on the operational details of take-back legislation implementations.

An aspect of product take-back that has not been modeled in detail to date is the process of product collection. Models usually include a per-unit collection cost, but the party or parties who are allowed to take EOL products back have not been explicitly modeled. In practice, one or more of the following parties may handle the collection: retailers, municipalities, producer-owned collection points, etc. The question of optimal reverse channel design has been studied in the absence of take-back legislation (e.g., Savaskan et al. 2004; Savaskan and Wassenhove 2006) and we believe that this question should be revisited for products facing take-back legislation. It is likely that the return volumes for products facing take-back laws will be higher, products returned may be in worse condition than products returned purely for economic reasons, etc. and because of these differences in the reverse channel, a channel structure that was optimal prior to legislation may not even be feasible under legislation.

To further the operations management research on product take-back legislation, we believe that it is necessary to better understand the recycling and remanufacturing processes and the related costs. Existing body of work on take-back laws have provided some insights regarding what type of legislation and compliance scheme is best for the environment and the producers, and the results, to a large extent, depend on how the compliance costs are calculated. Unfortunately, the functional forms that are used to estimate the costs of different steps in the collection and treatment processes are not validated by data but rather based on reasonable assumptions. For example, it is commonly assumed that collection cost is linear or convex in the number of units collected. Therefore, this line of research would benefit from testing the appropriateness of these assumptions by using the data that are available at the parties involved in different steps of compliance.

Finally, there is definitely room and need for empirical research in this field. Analytical models have provided some insights on what type of environmental benefits to expect from different types of legislation, and testing these findings empirically would make a nice research contribution.

References

Atasu A, Subramanian R (2009) Competition under product take-back laws: Individual or collective systems? Working Paper, College of Management, Georgia Institute of Technology, Atlanta

Atasu A, Van Wassenhove L (2010) Closed-loop Supply Chains: New Developments to Improve the Sustainability of Business Practices, chapter Environmental Legislation on Product Take-back and Recovery. CRC Press, Frozen

Atasu A, Van Wassenhove L (2011) An operations perspective on product take-back legislation for e-waste: Theory, practice and research needs. Technical report, Prod Oper Manag forthcoming

Atasu A, Guide D, Van Wassenhove LN (2008a) Product reuse economics in closed-loop supply chain research. Prod Oper Manag 17(5):483–497

Atasu A, Sarvary M, Van Wassenhove LN (2008b) Remanufacturing as a marketing strategy. Manag Sci 54(5):1731–1747

Atasu A, Sarvary M, Van Wassenhove LN (2009) Efficient take-back legislation. Prod Oper Manag 18(3):243–258

Atasu A, Guide D, Van Wassenhove LN (2010a) So what if remanufacturing cannibalizes my new product sales? Calif Manage Rev 52(2):56–76

Atasu A, Ozdemir O, Van Wassenhove L (2010b) Stakeholder perspectives on e-waste take-back legislation. Technical report, Georgia Institute of Technology, Atlanta

Clean Production Action (2007) How producer responsibility for product take-back can promote eco-design. Technical report, Tides Center, URL http://www.cleanproduction.org/pdf/cpa$_$ecodesign$_$Apr08.pdf

Debo LG, Toktay LB, Van Wassenhove LN (2005) Market segmentation and technology selection for remanufacturable products. Manag Sci 51(8):1193–1205

Dekker R, Fleischmann M, Inderfurth K, Van Wassenhove LN (2004) Quantitative Models for Closed-Loop Supply Chains. Springer Verlag, Germany

Dempsey M, Van Rossem C, Lifset R, Linnell J, Gregory J, Atasu A, Perry J, Sverkman A, Van Wassenhove LN, Therkelsen M, Sundberg V, Mayers K, Kalimo H (2010) Individual producer responsibility: A review of practical approaches to implementing individual producer responsibility for the WEEE Directive. Technical report, URL http://papers.ssrn.com/sol3/papers.cfm?abstract_id=1698695

Esenduran G, Kemahlioglu-Ziya E (2011) Complying with take-back legislation: A cost comparison and benefit analysis of three compliance schemes. Working Paper, Fisher College of Business, The Ohio State University, Ohio

Esenduran G, Kemahlioglu-Ziya E, Swaminathan JM (2011) The impact of take-back legislation on remanufacturing. Working Paper, Fisher College of Business, The Ohio State University, Ohio

Ferguson M (2009) Strategic issues in closed-loop supply chains with remanufacturing. In: Ferguson M, Souza G, (eds) Closed-Loop Supply Chains: New Developments to Improve the Sustainability of Business Practices. Taylor and Francis. Oxfordshire

Ferguson M, Toktay LB (2006). The effect of competition on recovery strategies. Prod Oper Manage 15(3):351–368

Ferguson M, Guide D, Koca E, Souza GC (2009) The value of quality grading in remanufacturing. Prod Oper Manage 10(3)

Ferrer G, Swaminathan J (2006) Managing new and remanufactured products. Manage Sci 52(1):15–26

Ferrer G, Swaminathan J (2009) Managing new and differentiated remanufactured products. Eur J Oper Res page Forthcoming

Ferrer G, Whybark DC (2001) Material planning for a remanufacturing facility. Prod Oper Manag 10:112–124

Galbreth MR Blackburn JD (2006) Optimal acquisition and sorting policies for remanufacturing. Prod Oper Manag 15(3)

Galbreth MR Blackburn JD (2010) Optimal acquisition quantities in remanufacturing with condition uncertainty. Prod Oper Manag 19(1)

Gray C, Charter M (2007) Remanufacturing and product design: Designing for the 7th generation. Technical report, The Center for Sustainable Design, Farnham

Groenevelt H Majumder P (2001) Procurement competition in remanufacturing. Working Paper, Simon School of Business, University of Rochester, Rochester

Gui L, Atasu A, Ergun O, Toktay B (2010) Fair and effcient implementation of collective extended producer responsibility legislation. Technical report, Georgia Insitute of Technology, Atlanta

Guide D Srivastava R (1997) Buffering from material recovery uncertainty in a recoverable manufacturing environment. J Oper Res Soc 48:519–529

Guide D, Teunter RH, Van Wassenhove LN (2003). Matching demand and supply to maximize profits from remanufacturing. Manuf Serv Oper Manag 5(4):303–316

Guide D, Souza G, van der Laan E (2005) Performance of static priority rules for shared facilities in a remanufacturing shop with disassembly and reassembly. Eur J Oper Res 164:341–353

HECC (2005) Electronic waste: An examination of current activity, implications for environmental stewardship, and the proper federal role. Technical report, house of committee on energy and commerce, URL http://www.access.gpo.gov/congress/house/pdf/109hrg/22988.pdf

Heese HS, Cattani K, Ferrer G, Gilland W, Roth AV (2005) Competitive advantage through take-back of used products. Eur J Oper Res 164:143–157

Inderfurth K, Flapper SDP, Lambert AJD, Pappis CP, Voutsinas TG (2004) Production planning for product recovery management in reverse logistics. Springer Verlag, Germany

IPR Works Press Release (2007) Alliance of NGOs and companies express strong criticism on electronic waste report. http://www.iprworks.org/documents/Press%20Release%20IPR%20Alliance%2014% 20November.pdf

Kekre S, Rao US, Swaminathan JM, Zhang J (2003) Reconfiguring a remanufacturing line at visteon, mexico. Interfaces 33(6):30–43

Majumder P, Groenevelt H (2001) Competition in remanufacturing. Prod Oper Manage 10(2):125–141

Nagurney A, Toyasaki F (2005) Reverse supply chain management and electronic waste recycling: a multitiered network equilibrium framework for e-cycling. Transportation Res Part E 41:1–28

Plambeck EL, Wang Q (2009) Effects of e-waste regulation on new product introduction. Manag Sci 55:333–347

Ray S, Boyaci T, Aras N (2005) Optimal prices and trade-in rebates for durable, remanufacturable products. Manuf Serv Oper Manag 7(3):208–228

Sander K, Schilling S, Tojo N, Van Rossem C, Vernon J, George C (2007) The producer responsibility principle of the WEEE Directive. Technical report, URL http://ec.europa.eu/environment/waste/weee/pdf/final_rep_okopol.pdf

Savaskan RC, Van Wassenhove LN (2006) Reverse channel design: The case of competing retailers. Manag Sci 52(1):1–14

Savaskan RC, Bhattachaya S, Van Wassenhove LN (2004) Closed loop supply chain models with product remanufacturing. Manag Sci 50(2):239–252

Sodhi MS Reimer B (2001) Models for recycling electronics end-of-life products. OR Spectrum, pages 97–115

Souza GC (2008) Closed-loop supply chains with remanufacturing. In: ZL Chen, R Raghavan (eds) Tutorials in Operations Research. INFORMS, Hanover

Souza GC Ketzenberg ME (2002) Two-stage make-to-order remanufacturing with service-level constraints. Int J Prod Res, 40(2)

Tojo N (2003a) EPR programmes: Individual versus collective responsibility. Technical report, The International Institute for Industrial Environmental Economics, Lund University, Sweden

Tojo N (2003b) EPR programmes: Individual vs. collective responsibility. Technical report, Industrial Institute for Industrial Environmental Economics, Lund University, Sweden

Toktay B, Wein L, Zenios S (2000) Inventory management for remanufacturable products. Manag Sci, 46(11):1412–1426

Toyasaki F, Boyaci T, Verter V (2010) An analysis of monopolistic and competitive take-back schemes for WEEE recycling. Production Operations Management, page Forthcoming

van der Laan E, Dekker R, Salomon M, Van Wassenhove L (1999) Inventory control in hybrid systems with remanufacturing. Manag Sci 45(5)

van der Laan E, Kiesmuller G, Kuik R, Vlachos D, Dekker R (2004) Stochastic inventory control for product recovery in reverse logistics. Springer Verlag, Germany

Van Rossem C (2008) Individual producer responsibility in the WEEE directive: From theory to practice? PhD thesis, the international institute for industrial environmental economics, Lund University, Sweden

Van Rossem C, Tojo N, Lindhqvist T (2006) Lost in transposition? A study of the implementation of individual producer responsibility in the WEEE Directive. Technical report, URL http://iprworks.org/documents/file/lost-in-transposition.pdf

Walther G, Schmid E, Spengler T (2008) Negotiation-based coordination in product recovery networks. Int J Prod Econ 111:334–350

Webster S, Mitra S (2007) Competitive strategy in remanufacturing and the impact of take-back laws. J Oper Manage 25(6)

Chapter 10
Product Disposition Decisions on Closed-Loop Supply Chains

Gilvan C. Souza

Abstract In this tutorial chapter we consider the problem of product disposition on a closed-loop supply chain: what should a firm do with a product return? Options include remanufacturing, materials recycling, and dismantling for spare parts. The answer to this question depends on the firm's objectives and environment. In many industries, remanufacturing is more profitable than dismantling for spare parts, which is more profitable than materials recycling. Recycling may or may not be profitable; and dismantling for spare parts may not be an option if the firm is not an Original Equipment Manufacturer (OEM). There may be uncertainty in demand for remanufactured products and/or the condition of returned products. We review here different decision support models for addressing this problem, and identify the conditions in which one approach is preferred over the other. We conclude with a quick review of the product (return) acquisition problem, which is related to the disposition problem, and that concerns the appropriate number of product returns to acquire in order to support the firm's closed-loop supply chain operations.

Keywords Remanufacturing · Product disposition · Closed-loop supply chains · Production planning

10.1 Introduction

In closed-loop supply chains, there are flows of products post-consumer use, which we refer to as returns, back to manufacturers, suppliers, or third party entities (i.e., not associated with the forward supply chain). These flows are in addition to the forward flows, from suppliers to manufacturers to distributors to consumers, prevalent in most supply chains. Returns can be classified into (i) end-of-life returns, which are obsolete products that have no further value to consumers; (ii) end-of-use returns, which are products with (potentially) some value left, after a period of use with consumers such as a leasing period; and (iii) consumer returns, which are products lightly used by consumers, and a result of lenient return policies by retailers. Upon receiving a return, a firm, which could be an Original Equipment Manufacturer (OEM), or a

G. C. Souza (✉)
Kelley School of Business, Indiana University, 47405 Bloomington, IN, USA
e-mail: gsouza@indiana.edu

T. Boone et al. (eds.), *Sustainable Supply Chains*, International Series in Operations Research & Management Science 174,
DOI 10.1007/978-1-4419-6105-1_10,

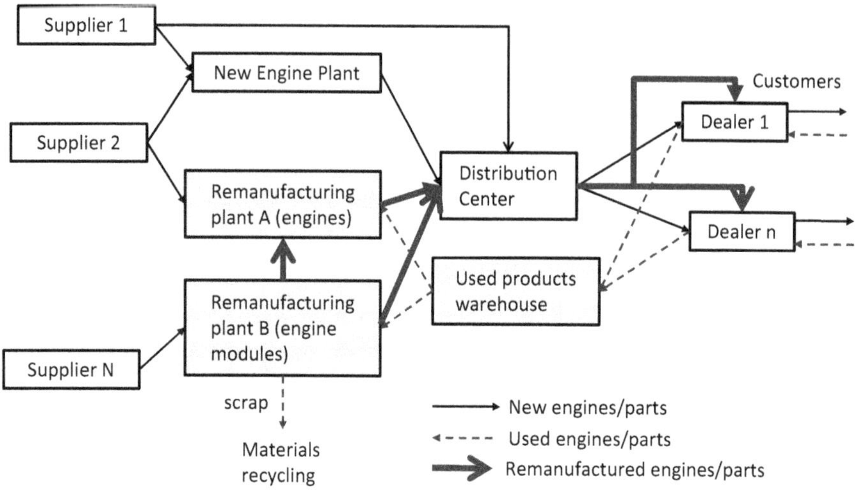

Fig. 10.1 Closed-loop supply chain for diesel engines (simplified)

third-party processor, makes a *disposition* decision. Disposition decisions include incineration (for energy recovery), recycling (for materials recovery), dismantling for spare parts, and remanufacturing. Remanufacturing is a value-added operation that restores a used product to a common operating and aesthetic standard, and where the core geometry of the product is preserved.

Because the loop of materials and products is closed, a closed-loop supply chain (CLSC) is critical to a firm's sustainability. Recycling provides a steadier source of raw materials, and avoids landfilling. Because the amount of energy and raw materials necessary for remanufacturing is a fraction of the amount necessary for the production of a new product with non-used content, remanufacturing is clearly beneficial from an environmental standpoint (Hauser and Lund 2003), although there are exceptions. For example, remanufacturing of certain appliances such as refrigerators is not environmentally optimal, because appliances consume about 90% of their energy, from a life cycle perspective, during the usage phase with consumers. As a result, replacing (inefficient) older refrigerators with modern, efficient ones based on newer technology is better for the environment; in this case materials recycling is optimal. However, in many cases remanufacturing is still the most sustainable disposition option, as indicated in numerous Life Cycle Assessment (LCA) studies.

An example of a CLSC is that of diesel engines and parts for Cummins, an OEM headquartered in Columbus, Indiana, which is shown in a simplified manner in Fig. 10.1. New engines are manufactured and shipped to a central distribution center for further distribution to thousands of dealers. In addition to new diesel engines and parts, which do not contain any remanufactured content, Cummins also sells remanufactured diesel engines and parts at a discount (relative to the corresponding new engine or part) of about 35%. Used engines or parts, the key inputs to remanufacturing, are obtained by Cummins as a result of customer trade-ins upon purchase of a new or remanufactured product; they are shipped from dealers to consolidation points, and then to a central used products depot. Used engines or parts are then shipped to one of two plants: engine remanufacturing (plant A), or part (or module) remanufacturing (plant B). For a diesel engine or part, remanufacturing consists of six different steps: (i) full disassembly; (ii) cleaning (through chemical bathing, sand blasting, and other methods); (iii) making a disposition decision for each part (keep for remanufacturing or dispose the part for materials recycling); (iv) remanufacturing the part; (v) re-assembly; (vi) testing. Remanufactured engines are shipped from plant A to the main distribution center, joining new engines or parts for distribution to dealers. Remanufactured parts or modules are shipped from plant B to either the distribution center, or to the engine remanufacturing plant A, depending on forecasts and current needs. Used parts not suited for remanufacturing are sold to recyclers. The flows depicted in Fig. 10.1 are simplified, but they convey the major flows in this CLSC. Used products are typically referred to as cores in industry.

In this chapter we focus on the optimal disposition decision for a firm in a closed-loop supply chain that processes used products; the firm could be an OEM (as Cummins) or a third-party remanufacturer. The chapter is written in a "tutorial" format, where we introduce a first (basic) decision model, and then add refinements. We do not provide a comprehensive review the literature in product disposition; to that end a state-of-the-art review is found in Fleischmann et al. (2010).

The available disposition decisions we consider here are: remanufacturing, dismantling for spare parts, and materials recycling. Although not exhaustive, this set is representative of many actual CLSCs, according to our experience. In many industries, remanufacturing is more profitable than dismantling for spare parts, which is more profitable than materials recycling. Recycling may or may not be profitable; and dismantling for spare parts may not be an option if the firm is not an Original Equipment Manufacturer (OEM), or if parts are subject to significant wear and tear during use. For the Cummins example above, the only two disposition decisions are remanufacturing and recycling, with remanufacturing being largely preferable to recycling, and recycling only occurring for products or parts not fit (or economical) for remanufacturing. For IBM, Pitney Bowes, HP, and other firms where products have electronic components not subject to significant wear and tear, dismantling for spare parts is another common disposition decision.

To make a decision into the appropriate disposition decision, the firm has to take into account several factors: time-varying demand forecasts for remanufactured products, demand for different spare parts, time-varying forecast of product returns, condition of product returns, profit margins of the different disposition alternatives, inventory holding costs, and penalty costs for unmet or backlogged demand for spare parts or remanufactured products. In particular, remanufacturing cost—and hence the profit margin of the remanufacturing disposition option—depends on the condition (quality) of the return used in remanufacturing.

This chapter is organized as follows. In Sect. 10.2, we introduce a relatively simple linear programming model for product disposition with two options: remanufacturing and salvaging (which could in practice mean dismantling for spare parts, or recycling). This model assumes an accurate forecast of product returns, along with their qualities, and is appropriate for a firm that leases its production with recovery at the end of the lease for remanufacturing. In Sect. 10.3, we introduce uncertainty in the quality of product returns, which results in a stochastic program. In Sect. 10.4, we introduce uncertainty in both the quantity of product returns, as well as demands for the various disposition options; the model however requires some simplifications such as unlimited capacity and a single quality grade for product returns. All previous models consider the case where the firm has an incoming return stream with little control over its quantity; in Sect. 10.5 we comment on the product acquisition decision—where the firm can actively procure a sufficient number of product returns to support its remanufacturing operations.

10.2 Base Linear Programming Model: Stable Demand, Return Quantity, and Quality

In this section, we present a basic model for disposition decisions, based on a OEM that leases its products. We consider a leasing scenario here because in leasing the firm has a good forecast of product returns per period, based on leasing expiration dates. Upon the end of a leasing period, a customer returns the used product to the firm. In this model, there are two possible disposition decisions for returns: remanufacturing, and salvaging. Salvaging here could mean dismantling for parts, or materials recycling, however, we do not model further details of the salvaging option, and instead simply consider a *given* unit profit margin associated with the salvaging disposition option. In each period, the firm receives a total of B^t end-of-lease returns, also called *cores*. Again, the value of B^t can be forecasted with a reasonable degree of accuracy. Cores are then graded and grouped into I different quality grades. A core of higher quality costs less to remanufacture (and consumes less remanufacturing capacity) than a core of lower quality. The proportion of each quality grade i in the mix of returned products received is r_i, which, for the time being, we assume to be known based on previous experience. Thus, the amount of cores of quality i in period t is given by $r_i B^t$. We consider a single product type; the model formulation is easily extended into multiple product types. One may view this problem as an aggregate production planning problem. Graded cores can be remanufactured to meet demand for remanufactured products. The demand for remanufactured products is forecasted over a planning horizon of T periods; in each period the forecast is D^t. Remanufactured products are identical, however, remanufacturing costs depend on the quality grade of the core used. Graded cores can be kept in stock to be remanufactured in the future. Graded cores can also be salvaged at any time; higher quality cores have higher salvage values corresponding to the opportunity to harvest higher quality cores for spare parts. The problem is to determine how many of the graded cores to be remanufactured, and how many of them to be salvaged each period. Due to capacity constraints, the firm may not be able to meet the non-stationary demand, as lower quality cores consume more capacity than higher quality cores. Therefore, backlogs are allowed. The objective is to maximize its total expected profit across a finite discrete-time horizon, $t = 1, \ldots, T$. The firm decides, for all quality grades, how many returns to remanufacture and how many to salvage (materials and parts), and in turn how many returns to hold for future periods. We summarize our notation below.

Indexes

i Quality category, $i = 1$ (best), …, I (worst)
t Time period, $t = 1, \ldots, T$

Parameters

D^t Demand for remanufactured products in period t
B^t Quantity of returns in period t
r_i Fraction of quality-i cores
C^t Available remanufacturing capacity in period t
k_i Unit remanufacturing capacity usage for quality-i cores
P_r Unit selling price for a remanufactured product
c_i Unit remanufacturing cost for quality-i returns
s_i Unit salvage value for quality-i returns
h_i Unit holding cost for quality-i returns
h_r Unit holding cost for a remanufactured product
h Unit inventory holding cost for un-graded cores
π Unit backlogging cost per period

Decision Variables

z_{it} Quantity of quality-i returns remanufactured in period t
v_{it} Quantity of quality-i returns salvaged at end of period t

Auxiliary Variables

u_{it} Inventory of quality-i returns at end of period t
y_t^+ Inventory of remanufactured products at end of period t
y_t^- Backlog of remanufactured products at end of period t

Our problem of finding the appropriate disposition decision (remanufacturing or salvaging) for returns in each period of the planning horizon can be formulated as a linear program as follows.

$$\max_{z_{it},v_{it}} \Pi = \sum_{t=1}^{T} \left\{ \sum_{i=1}^{I} ((P_r - c_i)z_{it} + s_i v_{it} - h_i u_{it}) - h_r y_t^+ - \pi y_t^- \right\} \tag{10.1}$$

Subject to

$$(y_{t-1}^+ - y_{t-1}^-) - y_t^+ + y_t^- + \sum_{i=1}^{I} z_{it} = D^t, \quad t = 1, \ldots, T \tag{10.2}$$

$$z_{it} + u_{it} - u_{i,t-1} + v_{it} = r_i B^t, \quad i = 1, \ldots, I; \quad t = 1, \ldots, T \tag{10.3}$$

$$\sum_{i=1}^{I} k_i z_{it} \le C^t, \quad t = 1, \ldots, T \tag{10.4}$$

$$y_t^+, y_t^-, z_{it}, v_{it}, u_{it} \ge 0, \quad i = 1, \ldots, I; \quad t = 1, \ldots, T. \tag{10.5}$$

The objection function (10.1) maximizes profit over the planning horizon, comprised of revenue from remanufacturing, revenue from salvaging, minus holding cost for graded cores, holding cost for remanufactured products, and penalty cost for backlogged demand of remanufactured products. Constraint (10.2) represents demand balancing constraints, (10.3) represents core balancing constraints, (10.4) represents remanufacturing capacity constraints, (10.5) are the non-negativity constraints.

Using this formulation, Ferguson et al. (2009) study the benefits of implementing a grading operation for categorizing product returns into different quality grades, as opposed to the case where all returns go into the same inventory "pile". They compute additional profits gained by implementing a grading scheme with I grades, where $I = 1$ means no grading (all returns go to the same pile and are remanufactured on a first-come-first-serve basis), $I = 2$ means a simple categorization (good, bad), and so forth. They find a concave relationship between profit improvement and number of grades used in the grading scheme, so that significant profit improvements can be achieved with three grades (good, medium, bad), and there is very little gain of implementing a grading scheme with more than five grades. Profit improvements of implementing a grading scheme are in the order of 5% or so, depending on a key parameter: how much returns exceed demand for remanufactured products, on average, per period. As the ratio of average returns to average demand for remanufactured products increase, the benefits of grading increase because the firm can be "pickier" when selecting returns to remanufacture, incurring lower remanufacturing costs, and less consumption of remanufacturing capacity, which allows the firm to carry less inventory of remanufactured products to protect against peak periods.

The problem (10.1–10.5) basically allocates returns to remanufacturing or salvaging, based on known returns (B^t) and demands (D^t) over the planning horizon. As discussed earlier, the assumption of known returns is a reasonable one for a leasing scenario. The assumption of known and deterministic demands is not as defensible, however, they can be dealt with by solving the problem on a rolling horizon, typical of aggregate production planning approaches in forward chains. One assumption that significantly simplifies the problem is that of known distribution of return qualities (i.e., r_i's are known and deterministic). We relax this assumption in the next section.

10.3 Stochastic Programming Approach: Uncertain Returns Quality

Now, consider the case where the total amount of cores is known in advance, however, their quality levels are uncertain. Thus, the number of cores of quality i in each period is uncertain, after grading. Relating to our previous model, we now, essentially, assume that the grading outcome r_i is a random variable. Outcomes should be thought of in terms of higher-level events (e.g., optimistic, average, pessimistic), to allow computational feasibility. Cores can be kept in stock to be graded in the future and graded cores can be kept in stock to be remanufactured in the future. The problem is to determine how many of the available cores to grade, how many of the

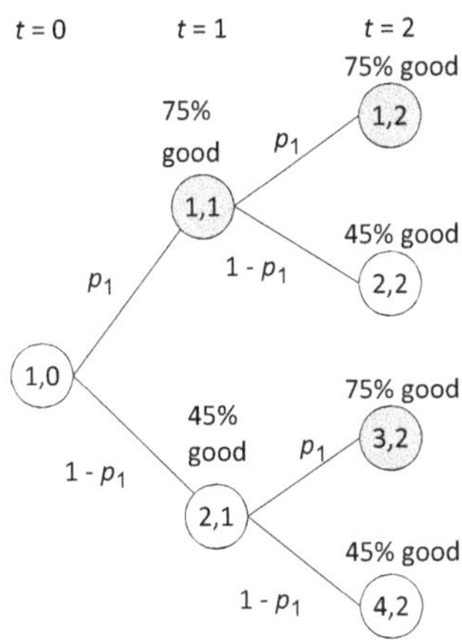

Fig. 10.2 Construction of scenarios (j, t)

graded cores to be remanufactured, and how many of them to be salvaged each period (say, dismantling for spare parts). This problem has been addressed by Denizel et al. (2010), who has formulated it as a multi-stage stochastic program, and we describe their model here. To illustrate the additional complications of uncertain grading outcomes, suppose there are two quality grades (good and bad), and two possible grading outcomes. In outcome 1, which happens with probability p_1, the grading process results in 75% of good quality cores, whereas in outcome 2, the grading process results in 45% of good quality cores. We now need to define $J(t)$ different *scenarios* for each time period t, where each scenario corresponds to a possible realization of grading outcomes in time t *and* previous periods. For example, period 2 has four possible scenarios ($J(2) = 4$) pertaining to the percentage of good quality cores in periods 1 and 2 ((0.75, 0.75), (0.75, 0.45), (0.45, 0.75), (0.45, 0.45)). We denote scenario j in period t as scenario (j, t). This is shown in Fig. 10.2. For a given time period t, define $(\text{pred}(j), t - 1)$ as the scenario, in period $t - 1$, that is the predecessor of scenario (j, t). For example, in Fig. 10.2, pred(1,2) = pred(2,2) = (1,1).

First, the amount of cores to be graded x is determined; based on that decision, the amounts of cores in each quality grade are revealed. These determine the scenarios in each period. Then for each scenario, the amounts of cores to remanufacture z, salvage and hold in inventory from each quality grade are decided upon. Additional notation is defined below.

(j, t) Scenario j of time period t, $j = 1, \ldots, J(t)$; $J(0) = 1$

Parameters

g Unit grading cost
$p^{(j,t)}$ Probability of scenario (j, t)
$r_i^{(j,t)}$ Fraction of quality-i cores under scenario (j, t)

Decision Variables

$x_t^{(j,t-1)}$ Amount of cores graded at time t under scenario $(j, t - 1)$
$z_i^{(j,t)}$ Quantity of quality-i cores remanufactured under scenario (j, t)

$v_i^{(j,t)}$ Quantity of quality-i cores salvaged at end of period under scenario (j, t)

Auxiliary Variables

$u_i^{(j,t)}$ Inventory of quality-i cores at end of period under scenario (j, t)
$y^{(j,t)+}$ Inventory of remanufactured products at end of period under scenario (j, t)
$y^{(j,t)-}$ Backlog of remanufactured products at end of period under scenario (j, t)
$b_t^{(j,t-1)}$ Amount of un-graded cores at the end of period t under scenario $(j, t-1)$

Our problem of finding the appropriate disposition decision for all periods in the planning horizon, under each probabilistic scenario (assuming zero initial inventories) can thus be formulated as:

$$\max \Pi = \sum_{t=1}^{T} \sum_{j=1}^{J(t)} p^{(j,t)} \left\{ \sum_{i=1}^{I} ((P_r - c_i) z_i^{(j,t)} + s_i v_i^{(j,t)} - h_i u_i^{(j,t)}) - h_r y^{(j,t)+} - \pi y^{(j,t)-} \right\}$$
$$- \sum_{t=1}^{T} \sum_{j=1}^{J(t-1)} p^{(j,t)} (g x_t^{(j,t-1)} + h b_t^{(j,t-1)}) \tag{10.6}$$

Subject to

$$y^{(\text{pred}(j),t-1)+} - y^{(\text{pred}(j),t-1)-} - y^{(j,t)+} + y^{(j,t)-} + \sum_{i=1}^{I} z_i^{(j,t)} = D^t, \quad \forall (j,t) \tag{10.7}$$

$$b_t^{(j,t-1)} + x_t^{(j,t-1)} - b_{t-1}^{(\text{pred}(j),t-2)} = B^t, \forall t, j = 1, ..., J(t-1) \tag{10.8}$$

$$z_i^{(j,t)} + u_i^{(j,t)} - u_i^{(\text{pred}(j),t-1)} + v_i^{(j,t)} - r_i^{(j,t)} x_t^{(\text{pred}(j),t-1)} = 0, \quad \forall (j,t) \tag{10.9}$$

$$\sum_{i=1}^{I} k_i z_i^{(j,t)} \leq C^t, \quad \forall (j,t) \tag{10.10}$$

$$y^{(j,T)} = 0, \quad j = 1, \ldots, J(T) \tag{10.11}$$

$$y^{(j,t)+}, y^{(j,t)-}, z_i^{(j,t)}, v_i^{(j,t)}, u_i^{(j,t)}, x_t^{(j,t-1)}, b_t^{(j,t-1)} \geq 0, \quad \forall i, (j,t). \tag{10.12}$$

The objective function (10.6) maximizes total profit. Profit is comprised of revenue from remanufactured products, and salvage revenue minus holding cost for graded cores, holding cost for remanufactured products, backlogging cost, grading cost, and holding cost for ungraded cores. The set of constraints (10.7) are inventory balancing constraints for remanufactured products: for a given scenario (j, t), one can meet demand from the current period's production and starting inventory of remanufactured products, or demand can be backlogged. Note that all scenarios (j, t) with a common predecessor share the same starting inventory $y^{y(\text{pred}(j),t-1)+}$. Constraints (10.8) and (10.9) display inventory balance constraints for un-graded and graded cores, respectively. Note that in (10.9), all scenarios (j, t) that share the same predecessor have the same amount of graded cores $x_t^{(\text{pred}(j),t-1)}$. Constraint

(10.10) regards the capacity constraint for period t. Finally, constraint (10.11) ensures that the firm cannot produce in excess of demand over the planning horizon, which is necessary because the objective function maximizes profit, which is based on production.

The formulation above can be easily modified to accommodate the case where demand for remanufactured products is uncertain. To that end, we redefine an outcome in any period to be the joint realization of the random grading process and random demand in that period. For example, if there are two possible grading realizations (good and bad), and three possible demand realizations (high, medium, and low), then each period has $2 \times 3 = 6$ probabilistic outcomes. The only change in the above formulation is that D^t is replaced with $D^{(j,t)}$ in (10.7). This simple modeling change, however, results in a much larger problem, because the number of decision variables increases exponentially in the number of scenarios.

Denizel et al. are able to solve problems of realistic size within reasonable computation times. For a six-period planning horizon, three quality grades, and five different grading outcomes, the problem has 222,647 variables and 101,556 constraints, which can be solved in Cplex in about one minute. Through an extensive numerical study based on industry data, Denizel et al. find that a firm's profit is heavily influenced by the shape of the remanufacturing cost curve as it relates to the quality of the core (convex increasing, linear, or concave increasing in the lower quality of the core), the profit margin of the salvaging disposition option, and the cost of grading. Other model parameters (such as shape of demand curve, backlogging and penalty cost) are not as critical. They conclude that firms should take actions toward reducing grading costs (through technology such as electronic data logs, which register usage, found in many products), and increasing the margin of the salvaging option (through use for spare parts, for example, as opposed to materials recycling).

The models presented so far assume a deterministic product returns stream, so that returns can be forecasted for the entire planning horizon with a reasonable degree of accuracy. Again, this assumption is reasonable if returns originate from leasing operations. Although both stochastic returns and demands can be incorporated into the stochastic programming formulation above, this option is not practical due to computational times, at least for a multi-period planning horizon. In cases where there are multiple sources of significant uncertainty: returns, demand for remanufactured products, and demand for the salvaging option (spare parts), another modeling approach is necessary, which we present in the next section.

10.4 Uncertain Demands and Return Quantity

With multiple sources of uncertainty—demand for remanufactured products, demand for spare parts, and product returns—one needs another approach to solving the disposition problem; it should be based on decision "rules" of simpler structure than the solution provided by the mathematical programming approach of previous sections. To derive simpler decision rules, however, it is necessary to make simplifications—for example, to assume unlimited remanufacturing capacity, and to assume a single quality grade for product returns—so as to formulate the disposition decision problem as a Markov decision process.

We illustrate the model by Inderfurth et al. (2001), which is reasonably general. There are n possible disposition decisions for stochastic product returns B^t, received at period t. Again, product returns are not categorized into different quality grades. Each disposition option has associated with it a demand stream. For example, disposition option 1 could be remanufacturing (with associated demand stream being demand for remanufactured products), disposition option 2 could be dismantling (with associated demand stream being demand for a single spare part originated from dismantling), and so forth. Assume that the firm can also dispose (say, through material recycling) returns it does not want to assign to one of the diposition options, at a unit cost of d. Formally, disposition option j faces stochastic demand D_j^t at time t, has processing lead time of λ_j periods, unit processing cost c_j, and unit shortage cost π_j. Unit holding cost for returns is h, and holding cost for option-j processed unit is h_j. The firm then decides on the disposition quantities z_{jt}, and disposal quantity v_t at time t. Denoting by α the one-period discount factor, the firm solves the following Markov decision process:

$$TC = \min_{v_t, z_{1t}, \ldots, z_{nt}} E\left[\sum_{t=1}^{T} \alpha^{t-1}\left[dv_t + \sum_{j=1}^{n} c_j z_{jt} + h\left(u_{t-1} - v_t - \sum_{j=1}^{n} z_{jt} + B^t\right)\right.\right.$$
$$\left.\left. + \sum_{j=1}^{n} h_j\left(y_{j,t-1} + z_{j,t-\lambda_j} - D_j^t\right)^+ + \sum_{j=1}^{n} \pi_j\left(D_j^t - y_{j,t-1} - z_{j,t-\lambda_j}\right)^+\right]\right]. \tag{10.13}$$

In (10.13), the expectation is taken with respect to $B^t, D_1^t, \ldots, D_n^t$; u_t and y_{it} are, just like before, the ending inventory of returns and processed units for option j at time t, respectively; they satisfy balancing equations similar to (10.2) and (10.3): $y_{jt} = y_{j,t-1} + z_{j,t-\lambda_j} - D_j^t$, and $u_t = u_{t-1} - v_t - \sum_{j=1}^{n} z_{jt} + B^t$. Inderfurth et al. (2001) show that a heuristic that sets a base stock policy for returns and base stock policies for each of the reuse options perform well. (A base stock policy for an option j maintains inventory position of processed units constant at z_j^*, where inventory position is defined as inventory on-hand minus backorders plus in-processing, or in transit, inventory.) However, it is necessary to specify an allocation rule, in case returns are not sufficient to meet demand for all disposition options in a period. They show that if returns are allocated to different disposition options based on their mean demands, then the performance of their heuristic is reasonably good.

The formulation above is reasonably general, and suitable to several situations. The modeling approach above is not appropriate, however, to a situation where profitabilities across different disposition differ significantly, because it allocates (potentially scarce) returns based on mean demands as opposed to expected marginal profits across disposition options. It is also not appropriate to a situation where the dismantling option produces more than one spare part (each with its own demand stream), since a disposition option only has one probability distribution of demand associated with it in the formulation above. Ferguson et al. (2011) propose an approach to address these situations. They provide a formulation and solution structure for a one-period problem with two disposition options (remanufacturing and dismantling for spare parts), which we describe here. They also provide a solution structure for the multi-period problem in the special case where dismantling only results in one type of part. The problem with multiple parts originating from dismantling in a multi-period setting is significantly more complex, and has not yet been solved.

We here describe their one-period model. At the beginning of the period, the firm receives B returns, where B is a random variable with continuous probability density function (pdf) denoted by $f_B(\cdot)$, and continuous cumulative distribution function (cdf) denoted by $F_B(\cdot)$. The firm then decides upon the number of units to be remanufactured z_r, at a unit cost c_r, and the number of units to be dismantled z_d, at a unit cost c_d. Demand for remanufactured products, sold at a unit price P_r, is a random variable D_r with cdf denoted by F_r. Each return dismantled results in a_i parts of type i. Demand for spare parts of type i is a random variable D_i with mean μ_i and cdf F_i. Returns that are not remanufactured or dismantled, given by $B - z_r - z_d$, are disposed of at a cost normalized to zero. Remanufactured units that are not sold are salvaged at a unit value of s_r. Demand for part i that is not met is assessed a unit penalty cost π_i, which can be though of as a higher cost of obtaining the part through an alternative supplier. Salvage value for all excess parts is zero. They assume the interesting case where $s_r \leq c_r < P_r$, i.e. the unit salvage value does not exceed the unit remanufacturing cost, which is smaller than the unit sales price for remanufactured products. Likewise, they assume $c_d < \sum_i \pi_i a_i$, i.e. the unit dismantling cost is smaller than the sum of the penalty costs avoided by dismantling one return.

The firm maximizes its one-period expected profit II:

$$\max_{\substack{z_r + z_d \leq B \\ z_r, z_d \geq 0}} \quad \Pi = E\Big[(P_r - s_r)\min\{D_r, z_r\} + (s_r - c_r)z_r - \sum_i \pi_i (D_i - \min\{D_i, a_i z_d\})\Big] - c_d z_d. \tag{10.14}$$

Considering that $E[\min\{D, z\}] = z - \int_0^z F(u)du$, where $F(\cdot)$ is the cdf of D, (14) becomes:

$$\max_{\substack{z_r + z_d \leq B \\ z_r, z_d \geq 0}} \quad \Pi = (P_r - s_r)\left(z_r - \int_0^{z_r} F_r(u)du\right) + (s_r - c_r)z_r - \sum_i \left[\pi_i \mu_i - \pi_i \left(a_i z_d - \int_0^{a_i z_d} F_i(u)du\right)\right] - c_d z_d. \tag{10.15}$$

The optimal solution of this problem is described in Lemma 1 below:

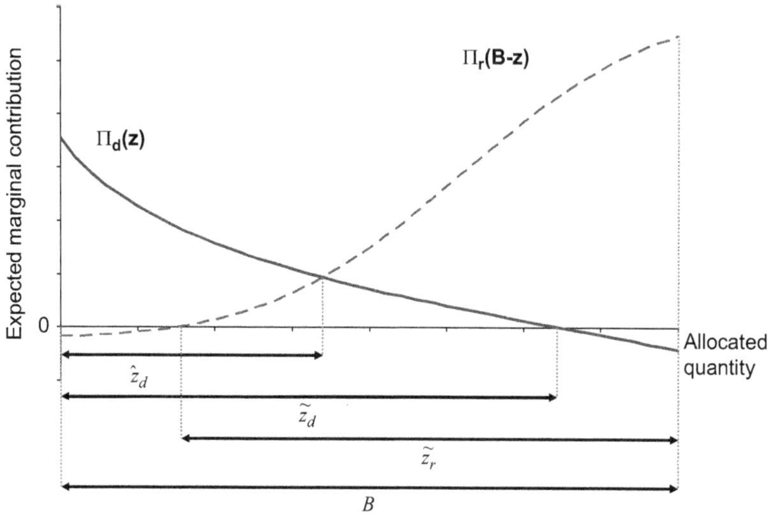

Fig. 10.3 Structure of the optimal disposition policy

Lemma 1 *Denote by*

$$\Pi_d(z) = \sum_i \pi_i a_i (1 - F_i(a_i z)) - c_d,$$

$$\Pi_r(z) = (P_r - v)(1 - F_r(z)) + s_r - c_r.$$

Then the optimal solution to the disposition problem (10.15) can be written as

$$(z_d^*, z_r^*) = \begin{cases} (\min\{B, \tilde{z}_d\}, 0) & \text{if } \Pi_d(B) > \Pi_r(0), \\ (0, \min\{B, \tilde{z}_r\}) & \text{if } \Pi_d(0) < \Pi_r(B), \\ (\min\{\tilde{z}_d, \hat{z}_d\}, \min\{\tilde{z}_r, B - \hat{z}_d\}) & \text{else,} \end{cases}$$

where $\tilde{z}_d$ and $\tilde{z}_r$ solve $\prod_d(z) = 0$ and $\prod_r(z) = 0$, respectively, and $\hat{z}_d$ solves $\prod_d(z) = \prod_r(B - z)$.

Lemma 1 postulates that if there are enough returns, the firm can satisfy demand for dismantling and remanufacturing at the levels $\tilde{z}_d$ and $\tilde{z}_r$, respectively, that set their respective marginal profits equal to zero. If there are not enough returns, however, the optimal dismantling quantity is set at a quantity that its marginal profit is equal to the marginal profit of remanufacturing.

Figure 10.3 depicts the optimal disposition policy, which is determined by the expected marginal contributions (and not by the respective mean demands, as in Inderfurth et al. (2001). The optimal decision is *not* a critical-level policy (where the quantity of returns allocated to the "cheaper" option equals the amount exceeding a certain threshold); this is because demand for both options are uncertain. In fact, if demand for parts is known and deterministic (and can be satisfied by dismantling d_d returns), then the optimal disposition decision is a critical-level policy:

Corollary 1 *Assume that* $D_i = d_d a_i$ *w.p. 1, for all i, and let* $\pi_d = \sum_i a_i \pi_i$. *Then the optimal remanufacturing and dismantling quantities satisfy:*

$$(z_d^*, z_r^*) = \begin{cases} (B - \min\{B, \check{z}_r\}, \min\{B, \check{z}_r\}) & \text{if } B \leq d_d + \check{z}_r, \\ (d_d, \min\{\tilde{z}_r, B - d_d\}) & \text{else,} \end{cases}$$

where $\check{z}_r = F_r^{-1}(1 - (c_r - s_r - \pi_d + c_d)/(P_r - s_r))$ *and* $\tilde{z}_r$ *is as defined in Lemma 1.*

When demand for spare parts is deterministic, the firm still determines the optimal amounts of remanufacturing and dismantling by comparing their respective expected marginal profits. The marginal profit of dismantling is constant and equal to $\pi_d - c_d$, for a quantity up to d_d. The quantity $\check{z}_r$ is the remanufacturing quantity for which the expected marginal profit equals the marginal dismantling profit. Note that $\check{z}_r$ is independent of B and is obtained as a Newsvendor solution. This is the amount of returns protected for the "high-margin customers", i.e. for remanufacturing, similarly to the two-class revenue management problem. Returns in excess of $\check{z}_r$ are available for dismantling, up to a maximum d_d. Any remaining returns should again be remanufactured, as long as the expected marginal profit remains positive.

Ferguson et al. (2011) provide a multi-period extension of this problem in the special case where there is only part type that originates from dismantling. Modeling multiple parts with very different demands introduces additional complexities, due to the interdependence between the dismantling decision and the different inventories levels of the different parts and due to the increasing state space dimensionality in the multi-period Markov decision problem.

10.5 Relationship with Product Acquisition

The decision models in this chapter focus on disposition decisions given an exogenous inflow of product returns (deterministic or stochastic) at a given quality. In many situations, firm can actually actively acquire product returns in the market place, and in some cases it can control the quality of returns acquired. The joint acquisition-disposition problem is fairly complex, but we can illustrate the acquisition problem with a single disposition option of remanufacturing. We illustrate two modeling approaches to the acquisition problem for remanufacturing. First, we illustrate the single-period model by Guide et al. (2003), which is motivated by remanufacturing of cell phones. There are N quality categories for returns; category i has remanufacturing cost c_i. To improve the quality of returns, the firm decides upon the acquisition price w_i for quality category i and realizes a corresponding return rate $B_i(w_i)$, where $B_i(\cdot)$ is increasing in w_i. Since the supply of returns, which is determined by w_i, constrains the amount of products that can be remanufactured, the firm should set (optimally) the price of remanufactured products P_r^* such that supply meets demand. The demand function (quantity sold) is $D(P_r)$, so the firm sets p_r^* such that $D(P_r^*) = \sum_{i=1}^N B_i(w_i)$, or $P_r = D^{-1}\left(\sum_{i=1}^N B_i(w_i)\right)$. The firm then decides on the acquisition prices w_i for each quality category i (which determines acquisition quantities) that optimizes its profit:

$$\max_{w_1,\ldots,w_N} \Pi(w_1,\ldots,w_N) = D^{-1}\left(\sum_{i=1}^{N} B_i(w_i)\right)\sum_{i=1}^{N} B_i(w_i) - \sum_{i=1}^{N} B_i(w_i)(w_i + c_i). \tag{10.16}$$

The solution to (10.16) depends on the shapes of the curves $B_i(w_i)$, and in general cannot be determined in closed form; Guide et al. (2003) provide an algorithm to find the optimal acquisition prices numerically. In the single-period model by Galbreth and Blackburn (2006), which is motivated by remanufacturing of toner cartridges, the acquisition price per unit w is exogenous and fixed, and demand D is known. Total remanufacturing cost, excluding acquisition cost for product returns, decreases in the total quantity of returns acquired B, which is the decision variable. This is because by sourcing more returns than demand $B > D$, the firm can select only the best D returns for remanufacturing. The firm thus finds B that minimizes the total cost $TC(B) = wB + B\int_{x=0}^{c^U} xg(x)dx$, where $g(x)$ is the probability density function for remanufacturing cost (with cdf $G(\cdot)$), $c^U = G^{-1}(\alpha)$, and $\alpha = \frac{D}{B}$. The value of $\alpha = \frac{D}{B}$ describes the acquisition policy: the firm remanufactures the top α fraction of returns it acquires. The trade-off is, the higher the total amount of returns collected B (lower α), the higher the acquisition cost but the total remanufacturing cost is lower because there are more returns of better quality in the pool of acquired returns. They find that the optimal policy B^* can be described by an α^* that is independent of B. That means, a single acquisition and grading policy is optimal for any value of D. Finding the optimal α^* can be done numerically through a simple line search. Galbreth and Blackburn (2006) model the distribution of returns quality (i.e., cost) through a known probability distribution ($G(\cdot)$), which is assumed to be invariant with respect to the quantity of returns acquired. This assumption holds if the acquisition quantities are large enough such that the probability distribution associated with the sample closely matches the overall population of returns. Galbreth and Blackburn (2010) extend their (2006) model to the case where there is uncertainty in the condition (i.e., the actual distribution of cost within the sample of products acquired) of used products. The 2010 model is thus appropriate for cases where the acquisition quantities are smaller, or where the firm has more limited information into the distribution of quality of used products in the field. They provide a closed form expression for the optimal acquisition quantities, which increases with the square root of the degree of condition variability.

There are other product acquisition approaches, and the reader is again referred to Fleischmann et al. (2010) for a complete review.

10.6 Conclusions

In this chapter, we have provided a tutorial and decision models for the optimal disposition decision in closed-loop supply chains. The disposition decision is: what should a firm do with a product return (a product post-consumer use)? The most common disposition decisions include remanufacturing, dismantling for parts, and recycling, which have been addressed in this chapter.

The appropriate management of closed-loop supply chains is an important sustainability component, because remanufacturing and recycling have been shown to significantly reduce environmental impact of supply chains by closing the loop of flows. It is often mentioned that sustainability means a focus on the 3P's of profit, planet and people. The models in this chapter focus on profit maximization, because an operation is not sustainable if it is not profitable. Our interactions with managers in many firms has indicated that most firms will not engage in environmentally friendly operations practices if they are unprofitable.

We have indicated that the optimal disposition decision is complex, because it involves the interaction of various random variables: quantity of returns per period, their qualities, demand for remanufactured products, and demand for various spare parts that can be obtained through dismantling. As a result, research has focused on certain scenarios where some of these random variables can be considered to be deterministic, for example, when the total number of incoming returns can be accurately forecasted through the planning horizon due to leasing agreements. Or, when there are no remanufacturing capacity constraints. The disposition problem that involves a dismantling option is particularly difficult. If there are multiple parts that can be obtained through dismantling, the problem has been solved for the single period case. For the multi-period case, Ferguson et al. (2011) have solved it for the case of a single part; for multiple parts in a multi-period situation, the problem remains unsolved.

We have also indicated the relationship between product acquisition and disposition. The models described in Sect 10.2, 10.3, and 10.4 concern a situation where the firm receives a stream of product returns (e.g., through leasing, consumer returns, etc.), and then must make the disposition decision that is most profitable. Product acquisition is important if the firm can actively procure product returns in the market place, such as in the case of cell phones, and toner cartridges.

References

Denizel M, Ferguson M, Souza GC (2010) Multi-period remanufacturing planning with uncertain quality of inputs. IEEE Trans Eng Manage 57(3):394–404

Ferguson M, Guide VD Jr, Koca E, Souza GC (2009) The value of quality grading in remanufacturing. Prod Oper Manage 18(3):300–314

Ferguson M, Fleischmann M, Souza GC (2011) A profit-maximizing approach to disposition decisions for product returns. Decision Sciences 42(3):773–798

Fleischmann M, Galbreth M, Tagaras G (2010) Production acquition, grading, and disposition decisions. In Ferguson M, Souza G, (Eds) Closed-loop supply chains: new developments to improve the sustainability of business practices. CRC Press, Boca Raton

Galbreth M, Blackburn J (2006) Optimal acquisition and sorting policies for remanufacturing. Prod Oper Manage 15:384–392

Galbreth M, Blackburn J (2010) Optimal acquisition quantities in remanufacturing with condition uncertainty. Prod Oper Manage 19(1):61–69

Guide VDR Jr, Teunter R, Van Wassenhove LN (2003) Matching supply and demand to maximize profits from remanufacturing. Manuf Serv Oper Manage 5:303–316

Hauser W, Lund R (2003) The remanufacturing industry: anatomy of a Giant. Boston University, Boston

Inderfurth K, de Kok AG, Flapper SDP (2001) Product recovery in stochastic remanufacturing systems with multiple reuse options. Eur J Oper Res 133:130–152

Chapter 11
Product Labeling, Consumer Willingness to Pay, and the Supply Chain

Robert L. Hicks

Abstract In this chapter I focus on the nexus between the supply chain and what is known about consumer willingness to pay for social and eco-labels. I briefly discuss the market penetration of product labels and what this means for product supply chains. I then focus on what is currently known about how consumers value label information. I go on to discuss import design issues such as who is certifying the label and the amount and type of information presented on the product label.

11.1 Introduction

In this chapter I examine what is known about consumer demand for social and eco-labels (hereafter referred to as eco-labels) and the implications for product supply chains. Since eco-labeling provides information to consumers about credence attributes (for example whether produce is organically produced or not, whether a product is child labor-free, etc.) related to a good's production and these production practices are impossible to observe at the time of purchase, eco-labeling is seen as a way to tap consumer demand for *how* a product is produced by providing credible information about production practices and methods. Provided that there are enough eco-consumers having a high enough valuation of the credence good, producers may alter their methods of production to appeal to the eco-consumer. Therefore, the efficacy of an eco-labeling program rests with the nexus between how consumers value the credence good, how credible the information is regarding a good's production practices and methods, and how costly it is for the firm to produce the eco-labeled product and meet label requirements.

During the past 40 years there has been explosive growth in product labeling. For eco-labels alone, there are now more than 377 active eco-labels in the world covering many product categories. Recognition rates for two of the broadest categories of labels are quite high: 54% (organic) and 82% (fair-trade) (Following the footprints 2011). Eco-labels also span a variety of product types including durable goods (Car Labels: Mitchell 2011) and have been adopted by Walmart the world's largest retailer

R. L. Hicks (✉)
129 Morton Hall, Department of Economics, The College of William and Mary, Williamsburg 23185, VA, USA
e-mail: rob.hicks@wm.edu

T. Boone et al. (eds.), *Sustainable Supply Chains*,
International Series in Operations Research & Management Science 174,
DOI 10.1007/978-1-4419-6105-1_11,

(Rosenbloom 2009). This growth and market penetration has occurred in 33 years following the first widely used eco-label, Germany's Blue Angel (Big Room Inc. 2011). In the United States, demand for organic labeled products have grown by 20–30% per year. Furthermore, labeled products now cover over 18% of global managed forests, 17% of coffee production, 8% of global tea productions, 1% of cocoa sales, 20% of global banana export (iSeal Alliance 2010). A survey of big-box retailers in the United States, Canada, the United Kingdom, and Australia revealed 2,219 products with product labels making green claims. In just over 2 years, the number of labeled products increased by 78% (Terrachoice Marketing 2009). For consumers, 35% of surveyed respondents say they would pay more for environmentally friendly products (Mintel Market Research 2010). This is consistent with rapid market growth for food and beverage labeled products (24% from 2006–2008) and personal care products (18% during the same period) (Mintel Market Research 2010). As for social labels, a similar pattern of growth has occurred. According to the Fair Trade Labeling Organizations International, global sales of "fair trade" products exceed $ 2.3 billion and the average annual growth rates in Canada (65%) and the United States (47%) are quite high. As for market share, there is high variability across countries ranging from a low of 0.5% in the United States to 20% in the United Kingdom (Ronchi 2006; Richardson and Stahler 2007)

The theoretical literature[1] on the efficacy of labels has focused on information asymmetries between the consumer and the producer, yet in a complex supply chain setting product standards may inform downstream buyers of upstream practices that could provide cost savings in producing the good. I provide some qualitative evidence that this is happening. I am doubtful, however, that labels will provide cost savings in enough settings to justify widespread label adoption. Therefore, I discuss in depth consumer willingness to pay for labeled products as a fundamentally important feature that producers must appeal to when adopting any labeling standard.

In the remainder of this chapter, I discuss how supply chains have adapted for meeting labeling requirements and the benefits and costs- as perceived by producers- of participating in labeling programs. I focus on the types of information and certifications being achieved in certification processes involving complex supply chains. Finally I turn to how consumers perceive and value eco-labels. I discuss the types of consumer benefits from credence goods and how the importance of labeling credibility and content in effectively harnessing the power of consumer sovereignty for labeled products.

[1] See Mason; Kotchen (2006); Swallow and Sedjo (2000); Hamilton and Zilberman (2006); Mahenc (2008); Amacher et al. (2004); Bougherara et al. (2002); Dosi and Moretto (2001); Mattoo and Singh (1994); Basu et al. (2003).

11.2 Assymetric Information and the Supply Chain

Firms produce goods having process and product attributes. Process attributes, describing the good's production process or method, may at the time of purchase be completely hidden to the consumer. What the consumer can observe at the time of purchase- product attributes- inform the consumer of the typical considerations made by consumers (for example taste, texture, quality) yet these product attributes rarely inform the consumer of the actual production method employed in the good's production. Consequently, the firm possesses more information about the full suite of product characteristics than the consumer. This asymmetric information problem has long been a focus of the labeling literature and the focus has been primarily on the asymmetry between producers and consumers. In this literature, the producer has full information about the production process and method, and producing the labeled good requires increased production and certification costs. If it was actually cheaper to use the greener production practices before the label, the firm would have already done so.

Complex supply chains possess a far different reality to the simple case presented above. There may also be information asymmetries between upstream manufacturers and downstream buyers in a supply chain. For example, the final consumer supplier of beauty products may not possess full information for each ingredient as to whether the production of that ingredient meets an environmental or social concern of the supplier or the supplier's customers. Expanding the information asymmetry problems for production attributes greatly complicates the certification and monitoring of labeling programs but at the same time presents potential benefits to firms.

For example, recent surveys of firms show that there are mixed opinions about whether they are better or worse off with product labeling standards. For example, for firms in a recent survey (iSeal Alliance 2010) say, "standards make things 'easier', 'reduce cost', 'increase efficiency' and 'help with supply chain management'". Labeling standards also facilitate communication and information transfer within the supply chain. Some go so far as to credit labels for improving internal communications and reporting: "Standards are the easiest way to pass on advice to suppliers" and for international supply chains allow firms or operations to "speak the same language". Nearly half of respondents from the financial sector reported that standards are used to screen investments. Beyond the possibility of uncovering hidden costs savings in the supply chain, product labeling standards may provide industry-wide reputational benefits as one executive said "standards need to be industry-wide so they form the baseline and set minimum performance levels for us to rope in the cowboys." (iSeal Alliance 2010). There are examples of labeling standards leading to production cost savings because the label increases transparency between elements of the supply chain (Following the footprints 2011).

Yet, the same survey reveals that certification requirements can be burdensome and costly. For example, consider a carbon label. Assimilating knowledge about the supply chain and mapping activities to emissions costs 3 M on average thirty thousand dollars per certified product and this does not include any operating cost increases from changes in production practices (Following the footprints 2011). On average, each firm is participating in four different standards systems and most would not be willing to adopt additional standards. If anything, most would like to reduce the number of standards as "there are too many overlapping standards, it is too confusing" (iSeal Alliance 2010), making the process of standards harmonization desirable for business. For many products, adhering to labeling standards leads to increases in production costs (one in three firms reported this) and the increased costs are not offset by market benefits (iSeal Alliance 2010). In the only market-wide study examining the costs of meeting product standards in a pre and post standard setting, the evidence points to rather high costs for meeting the standard (Hicks and Schnier 2008).

11.3 Willingness to Pay for Labeled Products

What exactly do consumers value when purchasing a more expensive labeled product rather than an otherwise identical unlabeled product? Knowledge of a goods production method may yield two types of benefits to consumers. Consider a GMO product (labeled "GMO Free") and the potential types of benefits[2]. The consumer may believe- despite the best scientific evidence to the contrary- that GMO Free products may be healthier than products containing GMO. Therefore, the consumer may be motivated to choose a GMO Free labeled product to avoid perceived health problems. There are also potential public benefits or costs to purchasing a GMO Free product. Consumers may feel- irrespective of their thoughts on the personal health impact- that a GMO Free product is beneficial to the environment since consuming a GMO Free product avoids the problem of gene leakage associated with GMO crops. Conversely, choosing a GMO Free product may signal to a consumer that more fertilizers and herbicides were used to produce the product (an environmental bad) relative to the GMO product. Consequently, evaluating consumer behavior around label information and attributing this behavior to public or private benefits is a difficult task, but an area of very active research in environmental economics (see for example (Carlsson et al. 2007)).

Irrespective of the type (or mix) of benefit enjoyed by the consumer when purchasing a labeled product, the literature has shown that there are several key factors related to consumer willingness to pay. There are numerous papers investigating consumer willingness to pay for labeled products (and the price premium for the credence attribute) in a label setting. Tables 11.1 and 11.2 list the studies along with the consumer product being studied. Some notable patterns emerge when examining evidence for willingness to pay. Market based studies such as Teisl et al. (2002) show small price premia compared to studies using experimental or hypothetical methods. Many of these studies go beyond merely measuring the existence of price premia to investigate the role of certifier, information content, and other factors on consumer

[2] Rousu examines these various types of benefits Rousu et al. (2007).

Table 11.1 Studies investigating price premia eco-labeled products

Study	Product
Anderson and Hansen (2004)	Wood products
Anderson et al. (2005)	Wood products
Bjorner et al. (2004)	Household items
Louriero et al. (2002)	Food
Nimon and Beghin (1999)	Apparel
O'Brien and Teisl (2004)	Wood product
Ozanne and Vlosky (1997)	Wood products
Teisl et al. (2002)	Seafood
Wessells et al. (1999)	Seafood

Table 11.2 Studies investigating price premia for fair trade and social labeled products

Study	Product
Elliot and Freeman (2003)	General products
Becchetti and Rosati (2007)	General products
Hiscox and Smyth (2006)	Apparel
Kimeldorf et al. (2004)	Apparel
Milkman (2004)	Apparel
Hicks (2007)	Coffee
Basu and Hicks (2008)	Coffee
Basu and Hicks	Coffee

willingness to pay. Given the strong evidence of the existence of consumer price premia for labeled products, it is useful to consider how these other factors influence consumer behavior.

11.3.1 Credibility, Certifier, and Stringency

One of the most important factors for the existence of a price premium is the credibility of the label. This is closely related to who might be certifying the product. Based on the literature, there doesn't seem to be any preferred type of certifier (an NGO or governmental body), except that third-party independent certification is preferred by consumers. For example, Roe and Teisl (2007) examines how consumers consider the role of potential label certifiers- a government agency (the US Food and Drug Administration) versus other governmental agencies (for example the US Environmental Protection Agency and the Department of Agriculture) and NGO's such as the American Heart Association. Also, it is interesting that this paper examines the presence versus absence of GMO (one product is labeled "No GMO" while the other is labeled "GMO"). The degree of credibility varied with the information provided on the label. Of particular interest was the difference in providing a positive label ("GMO Free") versus a negative label ("Contains GMO"). The negative label was always viewed by consumers as being more credible than a positive label regardless of certifying agency. Also, simply supplying personal contact information for the certifier improved label credibility.

At the outset of a labeling program, the label certifier must decide on the degree of label stringency. This decision will likely impact producers as more stringent label requirements are normally associated with higher costs of supplying a labeled product vis-a-vis an unlabeled one. Too stringent a label and firms may leave a product market (under a mandatory labeling scheme) or choose to supply the unlabeled product in a voluntary label setting. From the consumer perspective, more stringent labels may be more valuable than weak labels as consumers will not be willing to pay much for products they perceive as being merely green-washed. For example, in the case of dolphin-safe tuna, the initial label requirements were quite stringent, requiring vessels to avoid dolphins for the duration of complete cruises often lasting as long as 90 days. This stringency has been attributed as being an important factor in the relocation of the tuna fleet to areas of the world where tuna dolphin interactions do not occur. In response, the label stringency was relaxed and tuna was labeled as dolphin safe on a day by day basis[3]. Ironically, this more relaxed label requirement placed *greater* record keeping requirements on the supply chain but significantly reduced the cost of label compliance.

The majority of programs are run by non-profit NGO's followed by a for-profit organization. A very small minority of programs (3%) were run by an industry association. Programs run by non-profits tended to be more closely associate with higher label stringency (more site visits, auditing) (Big Room Inc. and the World Resources Institute 2010) and most labels (64%) are third party verified rather than second-party (27%). Nearly all (92%) of the surveyed labels required certification (rather than registration only) (Big Room Inc. and the World Resources Institute 2010). Some products are using labels that purport to be certified by third parties and are not (Terrachoice Marketing 2009). It is also notable, that nearly all studied labels violated at least one recommendation from a list of best marketing practices set forth by Terrachoice marketing.

11.3.2 Information Content

Binary versus Performance Labels Perhaps the most importance feature of a label is the information linking the performance of the label (via the supply chain) to consumer demand. One example is performance-based labeling. Beyond merely reporting whether a product passes a labeling requirement or not (a binary label) some product labels include the underlying performance of the labeling program with respect to the environmental or social credence attribute the program purports to influence. One prominent example of a performance label is the the tiered LEAD program for building sustainability- Bronze, Silver, Gold, Platinum certification under the LEAD program (Big Room Inc. and the World Resources Institute 2010).

[3] The requirement was for a set by set basis, where the term 'set' refers to each time a vessel placed fishing gear in the water.

As a building receives higher certification levels, it is associated with better environmental performance with respect to energy use, the impact of construction materials, and environmental impacts of the building's lifecyle.

Hicks (2007) and Basu and Hicks (2008) examine consumer WTP with varying degrees of information concerning the performance of a coffee fair trade labeling program. They show that compared to a standard binary label, providing additional information about the proportion of program participants and the degree of revenues flowing back to participating farmers leads to a substantial increase in consumer willingness to pay relative to a binary fair trade labeled coffee. The findings indicate that if there are easily communicated and understood performance measures, then product labeling that includes performance attributes may lead to even more incentives for producers to switch to desirable production practices.

The issue of performance labeling and increasing levels of consumer demand, while potentially attractive for producers, also places more burden on producers to collect information and certify the processes used in the supply chain. In the case of fair trade coffee, knowledge of each level of the supply chain- from farmers, to the fair trade cooperative, buyers, importers, roasters, and retailers- is necessary in order to assure consumers of the performance of the labeling program. At present, not many of the world's eco-labels are ready to move to performance based labels and most are of the binary type. While most labels make public their certification criteria (87%) only 46% actively monitor or have studied the environmental impact of their ecolabeling regime (Big Room Inc. and the World Resources Institute 2010). As a recent UN study reports, "not only is there a lack of information on the environmental effectiveness of ecolabelling, but there is an almost complete lack of independent data on its trade and economic effects." (United Nations Environment Programme 2005).

While the information burden on producers can be quite high under a performance-based label, it may help to overcome the closely tied issues of credibility and certifying agency. Under performance labels, the importance of the certifying agency (whether third party, government agency, or industry group) is much less important for consumers when credible performance information is included with the label. There is also conflicting evidence regarding more detailed performance based information. For example, Roe et al. (1999) found cases where qualitative content trumped actual measured content information concerning nutrition labels for food. Whether a performance-based label is superior to the traditional binary label really depends on the knowledge of the consumer vis-a-vis the credence attribute being communicated by the label and the ease of measuring the performance attributes.

Specific versus General Labels The intent of most eco-labeling programs is to shift production towards greener less harmful methods of production, and many eco-labeling programs are quite specific as to the credence attribute being communicated by the label. What consumers may want to know is the overall environmental impact of their product purchase relative to an unlabeled product and the presence of a label (for example "Organic") may not sufficiently inform consumers.

This is particularly germane under recent mandatory carbon labels being introduced in Europe (in particular France's Grenelle 2 Law passed in 2010). While information about a product's carbon impact is a very important environmental credence attribute, it may not tell the whole environmental story. For example a product may score quite well with respect to its carbon output yet might be particularly bad for biodiversity or water quality impacts. Under Grenelle 2, France has decided that for some product categories, the carbon label must include other environmental data such as water use and biodiversity impacts (Following the footprints 2011). Having a broader label that includes an index over all environmental impacts has its appeal but may have some drawbacks as well. First, an index that aggregates over a broad set of environmental impacts may be very difficult for consumers to understand. Second, measuring an array of environmental impacts or performance data places increasing burdens on the supply chain, and plays into producers' concerns about label proliferation and burden (iSeal Alliance 2010). Despite these concerns, Walmart has recently moved toward developing an index label that describes impacts over a variety of environmental and social factors into a single measurable attribute (Rosenbloom 2009).

Limits of Information Better and more complete information may not always lead to large changes in consumer behavior. For example, Teisl et al. (2008) examines WTP for eco-labeled durable goods (for example cars). Perhaps the most interesting finding is that information on the environmental performance of cars did not induce people to switch classes (for example sedans versus SUVs) but did help people get greener within a product class. To change car classes, other large-scale changes need to occur, such as a fundamental change in gas prices. Absent that, the authors argue that an education program in conjunction with a product label is perhaps the only way to induce large changes in behavior.

Environmental information was also most influential in consumer choice when linked with private benefits. Bond et al. (2008) examines organic label claims on fresh produce and explores how information on private health benefits versus a label describing only public health benefits is considered by consumers. They find that the private benefits are more important to consumers than the organic label. However, there are also important interaction effects between the public and private benefits of an eco-label. When health and organic bundling occurs in a label there is more than a linearly additive effect with respect to the consumers willingness to pay for the labeled product. Consequently, from the perspective of designing a labeling program, there are benefits to identifying products where public and private benefits occur together and communicating this information on the product label.

11.4 Conclusion

In this chapter, I explore what is important for consumers when purchasing labeled products and reflect on what that means for a product's supply chain. While a richer set of information concerning a product's production is likely to lead to a more credible label and be easier for consumers to evaluate, it may also require a great deal of information from the product's supply chain. In that context I discuss what is currently known in the labeling literature about the role of certifier, the degree and content of information, and the limits of labels in influencing consumer willingness to pay for labeled products.

References

Amacher GS, Koskela E, Ollikainen M (2004) Environmental quality competition and eco-labeling. J Environ Econ Manage 47(2):284–306
Anderson RC, Hansen EN (2004) Determining consumer preferences for ecolabeled forest products: an experimental approach. J Forestry 102(4):28–32
Anderson RC, Laband DN, Hansen EN, Knowles CD (2005) Price premiums in the mist. For Prod J 55(6):19–22
Basu AK, Chau NH, Grote U (2003) Eco-labeling and stages of development. Rev Dev Econ 7(2):228–247
Basu AK, Hicks RL (2008) Label performance and the willingness to pay for fair trade coffee: a cross-national perspective. Int J Consum Stud 32(5):470–478
Basu AK, Hicks RL "Poverty Alleviation through Social Labeling Programs? Information Valuation and Willingness to Pay for Fair Trade Coffee" March 2011 Working Paper, Department of Economics, The College of William and Mary
Becchetti L, Rosati FC (2007) Global social preferences and the demand for socially responsible products: empirical evidence from a pilot study on fair trade consumers. The World Econ 30(5):807–836
Big Room Inc. and the World Resources Institute (2010) Global ecolabel monitor: towards transparency
Big Room Inc. Ecolabel Index Website May (2011) URL http://www.ecolabelindex.com/
Bjorner TB, Hansen LG, Russell CS (2004) Environmental labeling and consumers' choice-an empirical analysis of the effect of the nordic swan. J Environ Econ Manage 47(3):411–434
Bond CA, Thilmany DD, Bond JK (2008) What to choose? the value of label claims to fresh produce consumers. J Agric Resour Econ 33(3)
Bougherara D, Grolleau G, Lockeretz W (2002) Can ecolabeling mitigate market failures? an analysis applied to agro-food products. In Ecolabels and the greening of the food market, Boston, Massachusetts, USA, November 7–9, (2002). pp 111–119. Friedman School of Nutrition Science and Policy, Tufts University, 2002
Carlsson F, Frykblom P, Lagerkvist CJ (2007) Consumer benefits of labels and bans on gm foods-choice experiments with swedish consumers. Am J Agric Econ 89(1):152–161
Dosi C, Moretto M (2001) Is ecolabelling a reliable environmental policy measure? Environ Resour Econ 18(1):113–127
Elliott KA, Freeman RB (2003) Can labor standards improve under globalization? Peterson Institute
Following the footprints. The Economist, May, 2011
Hamilton SF, Zilberman D (2006) Green markets, eco-certification, and equilibrium fraud. J Environ Econ Manage 52(3):627–644
Hicks R (2007) Performance-based labeling. New frontiers in environmental and social labeling, pp 37–58
Hicks RL, Schnier KE (2008) Eco-labeling and dolphin avoidance: a dynamic model of tuna fishing in the eastern tropical pacific. J Environ Econ Manage 56(2):103–116
Hiscox MJ, Smyth NFB (2006) Is there consumer demand for improved labor standards? evidence from field experiments in social labeling. Department of Government, Harvard University
iSeal Alliance (2010) The iseal 100: a Survey of thought leader views on sustainability standards

Kimeldorf H, Meyer R, Robinson I, Prasad M (2004) Consumers of the world unite: a market-based response to sweatshops. Labor Stud J 29(3):57–79

Kotchen MJ (2006) Green markets and private provision of public goods. J Polit Econ 114(4):816

Loureiro ML, Jill JM, Ron CM (2002) "Will consumers pay a premium for eco-labeled apples?" J Consum Aff 36(2):203–219

Mahenc P (2008) Signaling the environmental performance of polluting products to green consumers. Int J Ind Organ 26(1):59–68

Mason CF (2011) "Eco-labeling and market equilibria with noisy certification tests." Environ Resour Econ 4(48):537–560

Mattoo A, Singh HV (1994) Eco-labelling: policy considerations. Kyklos 47(1):53–65

Milkman R (2004) Sales results from ucsb pants experiment. Unpublished paper. Department of Sociology, University of California, Santa Barbara

Mintel Market Research (2010) Are Americans willing to pay more green to get more green? March 2010. URL http://www.mintel.com/press-centre/press-releases/514/are-americans-willing-to-pay-more-green-to-get-more-green

Mitchell Josh (2011) Autos get new fuel labels. Wall Str J. Accessed 26 May 2011

Nimon W, Beghin J (1999) Are eco-labels valuable? evidence from the apparel industry. Am J Agric Econ 81(4):801–811

O'Brien KA, Teisl MF (2004) Eco-information and its effect on consumer values for environmentally certified forest products* 1. J Forest Econ 10(2):75–96

Ozanne LK, Vlosky RP (1997) Willingness to pay for environmentally certified wood products: a consumer perspective. For Prod J 47(6):39–48

Richardson M, Stahler R (2007) Fair trade. Working Paper 0709, University of Otago, Dunedin, New Zealand

Roe B, Levy AS, Derby BM (1999) The impact of health claims on consumer search and product evaluation outcomes: results from fda experimental data. J Public Policy Mark 18(1):89–105

Roe B, Teisl MF (2007) Genetically modified food labeling: the impacts of message and messenger on consumer perceptions of labels and products. Food Policy 32(1):49–66

Ronchi L (2006) Fairtrade and market failures in agricultural commodity markets. World Bank Policy Research Working Paper 4011, Washington D.C.

Rosenbloom S (2009) At wal-mart, labels to reflect green intent. New York Times. Accessed July 15 2009

Rousu M, Huffman WE, Shogren JF, Tegene A (2007) Effects and value of verifiable information in a controversial market: Evidence from lab auctions of genetically modified food. Econ Inq, 45(3):409–432

Swallow SK, Sedjo RA (2000) Eco-labeling consequences in general equilibrium: a graphical assessment. Land Econ 76(1):28–36

Teisl MF, Roe B, Hicks RL (2002) Can eco-labels tune a market? evidence from dolphin-safe labeling* 1. J Environ Econ Manage 43(3):339–359

Teisl MF, Rubin J, Noblet CL (2008) Non-dirty dancing? interactions between eco-labels and consumers. J Econ Psychol 29(2):140–159

Terrachoice Marketing. The seven sins of greenwashing. Environmental claims in consumer markets. April 2009

United Nations Environment Programme (2005) The trade and environmental effects of ecolabels: assessment and response. United Nations Environment Programme

Wessells CR, Johnston RJ, Donath H (1999) Assessing consumer preferences for ecolabeled seafood: the influence of species, certifier, and household attributes. Am J Agric Econ 81(5):1084–1089

Chapter 12
Carbon Foot Print of Products—Supply Chain Approach

Tonya Boone, Ram Ganeshan and Vaidyanathan Jayaraman

Abstract This Chapter explores the carbon footprinting process from a supply chain perspective. The overriding premise is that the footprinting exercise needs to encompass the product's entire life cycle from the extraction of raw material, to the manufacture, transport, and use of the product; ending eventually in the disposal and recovery of the product. This Chapter will also explore issues and challenges in (a) organizing for and establishing an accurate carbon footprint and (b) acting on it to have a lesser impact on the environment.

12.1 What is a Carbon Footprint?

A carbon footprint of a product or a service measures the total amount of carbon dioxide-equivalent (CO_2-eq.) gases that it produces throughout its life cycle. This typically includes the amount of CO_2-eq. produced in the extended supply chain: from the extraction of raw material, to the manufacture, transport, and use of the product; ending eventually in the disposal and recovery of the product. Other greenhouse gases—Methane, Nitrous Oxide, Sulfur Hexafluoride and others—are converted to its "CO_2 equivalent" weight so a single CO_2-eq. number can represent the environmental impact of the product. The primary objective of a footprint is to establish the impact of the product or service and take appropriate action to reduce this impact.

At the firm, and specifically at the supply chain level, there is an increased awareness of climate change as a major strategic priority. In addition to addressing the global problem of climate change, firms have multiple reasons to address greenhouse gas (GHG) emissions. First, customers and supply chain partners are now

T. Boone (✉) · R. Ganeshan
Mason School of Business, College of William and Mary, 23185 Williamsburg, VA, USA
e-mail: Tonya.Boone@mason.wm.edu

R. Ganeshan
e-mail: Ram.ganeshan@mason.wm.edu

V. Jayaraman
Department of Management, School of Business Administration,
University of Miami, 33124 Coral Gables, FL, USA
e-mail: vaidy@miami.edu

T. Boone et al. (eds.), *Sustainable Supply Chains*,
International Series in Operations Research & Management Science 174,
DOI 10.1007/978-1-4419-6105-1_12,

demanding an organized approach to carbon management. A firm may risk its reputation if it does not understand and act on the broader impact of its products and services—the embedded emissions of purchased goods and the downstream distribution, use, and disposal of its products. For example, Wal-Mart has indicated that it will reduce the GHG emissions by 20 million tons by 2015—a target that cannot be achieved without the full involvement of all its major stakeholders including suppliers and customers.

Many firms look at carbon management as an opportunity to mitigate risk and reduce possible disruptions in the supply chain. To maintain assurance of supply, for example, Starbucks Coffee Company initiated C.A.F.E. (Coffee and Farmer Equity) practices to evaluate, recognize, and reward producers of high-quality sustainably grown coffee. In the electronics sector, there is a move to follow RoHS and WEEE[1] guidelines even when not mandated by local laws. For example, in addition to significant reductions in the carbon footprint[2], the latest generation of Apple's products are Lead, Brominated Flame-Retardants, Polyvinyl chloride, and Mercury free.

Another driver for carbon management is both ecological and economic efficiency. Reducing the carbon footprint often translates to lower use of energy, reduction in waste streams, and using recycled material in the supply chain. Herman-Miller's Mirra chairs, for example, are manufactured at a LEED[3] Pioneer building and the production line utilizes 100% green power. No air or water emissions are released in the chair's production. The chair, made from steel, plastic, foam, and textile is 96% recyclable at the end of its life (approximately 12–15 years) and is currently made from 33% recycled materials.

Finally, firms are aggressively managing GHG emissions in anticipation of stricter laws that will curb emissions. For example, the European Union has the Emissions Trading Scheme (EU ETS) that functions as a "cap and trade" system. The 2013 cap is about 2.04 billion "allowances" (each allowance is the right to emit one ton of CO_2). These allowances will be allocated among member nations and each member state grants their companies allowances through what is called a national allocation plan (NAPs). The EU ETS covers power stations and other combustion plants, oil refineries, coke ovens, iron and steel plants and installations producing cement, glass, lime, bricks, ceramics, pulp, paper and board, bulk organic chemicals, hydrogen, ammonia and aluminum. Depending on their emissions, the ETS allows a firm to buy or sell allowances to meet specified targets. Installations that are non-complaint will be subject to stiff penalties. The cap will decrease each year by 1.74% of the average annual total quantity of allowances issued by the Member States in 2008–2012. While the US does not have a cap and trade scheme, there have been various attempts to introduce similar Bills and in all likelihood, climate legislation will be a

---

[1] EU legislation on the Restriction of Hazardous Substances (RoHS) and the Waste Electronic, Electrical, Equipment (WEEE).

[2] The carbon footprint of the 2010 160 GB iPod Classic is 50% of the footprint of the 2001 5 GB iPod Classic.

[3] Leadership in Energy and Environmental Design (LEED) is a third party green building certification developed by the U.S. Green Building Council (USGBC).

reality in the US in the near future. Such regulations may significantly impact supply chains and how firms view their trading partners. For example, potential taxes on energy or embedded carbon emissions in products and services may significantly impact the cost of purchased goods or components used in the manufacturing and production processes of a company. Measuring and managing these emissions help companies plan for such potential policy regulations and guide supplier selection and product design.

12.2 Tracking of a Supply Chain Carbon Footprint

The Greenhouse Gas Protocol (GHG Protocol) is the most widely used international accounting tool for businesses to measure and manage greenhouse gas emissions. Through a partnership between the World Resources Institute (WRI) and the World Business Council for Sustainable Development (WBCSD), the GHG Protocol sets standards for preparing the "lifecycle inventory," a mechanism via which carbon footprints can be measured, managed, and mitigated.

While there are many other protocols like ISO 14000 and UK's PSA 2050, they all have similar components. The GHG protocol[4], for example, specifies three "scopes" of emissions:

- **Scope 1:** Direct GHG emissions are from sources that are owned and operated by the company. This includes all "within firm" supply chain operations, including facilities.
- **Scope 2:** This accounts for GHG emissions arising from purchased energy.
- **Scope 3:** Includes all other indirect emissions related to the firm's supply chain. This can mean anything from employee travel, to embedded emissions in products purchased or processed by the firm, to "downstream" supply chain emissions such as distribution, retail, the use of the product, and the eventual recapture of the waste stream the product or service generates.

Firms typically report Scope 1 and 2 but often do not take detailed inventory of emissions of Scope 3 since one firm's Scope 3 are likely another firm's scope 1. There are also obvious difficulties measuring scope 3 since it is outside the firm boundary and often outside the firm's control. However, much of the product or service's GHG emissions are often in the supply chain (scope 3)—several orders of magnitude higher than firm-level emissions. Managing the embedded carbon in the supply chain is a more comprehensive approach and can help realize both economic and ecological efficiencies for the firms in the supply chain.

[4] We use GHG "Scopes" for illustration. Other protocols measure emissions similarly—PAS 2050 simply follows the product or service supply chain without formally designating scope.

12.2.1 Organizing for Footprint Analysis

Any carbon measurement and reduction process needs a "lead" or a "reporting" firm—a firm that owns the product or the service. For example, as the ensuing discussion will show, Pepsi Co. owns the footprinting process of Walkers crisps or Tropicana pure premium orange juice; Apple owns the process for the iPhone4; and Toyota owns the footprinting process of the Prius. The primary responsibility for the lead firm is to create an organization within the firm and related processes that will identify products and services to footprint, marshal data sources, create the footprint, have it verified, and finally engage supply chain partners for action. Typical participants in the footprinting project are divisions within the lead firm, suppliers, downstream distribution channels and customers, consulting firms, third-party accountants to verify footprint, research organizations and environmental watchdogs that help identify products and services for GHG reduction and audit and critique GHG mitigation efforts.

The governance structure for strategic sustainability efforts that sets overall goals and specific company-wide GHG reduction targets typically involves a cross-disciplinary committee or executive officers reporting to the Board of Directors (or a Board Committee).

Pepsi Co., for example, has formed the Environment Sustainability Leadership Team (ESLT) to address sustainability issues. The ESLT, which is led by Divisional Supply Chain chiefs, oversees the implementation of climate change mitigation in Pepsi Co.'s supply chains. The ESLT directly advises the CEO and the Board of Directors.

The ESLT's focus areas are energy, water, solid waste, packaging, and sustainable agriculture. The ESLT works globally with regional Pepsi Co. leaders to identify climate change mitigation opportunities and continuously assesses and implements mitigation strategies and processes.

Apple uses an "Executive Team" to addresses their sustainability strategy. The executive team is a group of the company's senior-most executives and regularly reviews each new product during its development, focusing on material and design choices, the supply chain, packaging, and product energy efficiency. They get inputs and advise from different Apple divisions on how best to devise a GHG emission mitigation strategy. According to Apple, they are the first technology company to publish the supply chain footprints of all their products.

Wal-Mart has organized its sustainability strategy around collaborative partnerships called "Sustainable Value Networks" (SVN). Their SVNs consist of Wal-Mart Associates, NGOs, government officials, academics and suppliers. Wal-Mart has SVNs dedicated to product categories, measuring climate impact, and supply chain efficiency.[5] Each network is lead by a "Network Captain" who is a Director or Vice-president level and guides the network towards meeting climate targets. Every

[5] Wal-Mart's SVNs are in Agriculture & Seafood, Textiles, Jewelry, Electronics, Wood and paper, Chemical-intensive products, Greenhouse Gas, Sustainable facilities, Logistics, Waste, and packaging.

quarter, networks report on their progress on their climate strategy through a network executive sponsor—typically at a senior vice-president level or higher—to the CEO.

An effective governance structure specifies the processes both at the strategic level as well as the operational level by which GHG reductions can be targeted. At Wal-Mart, for example, as part of the strategic directive to reduce 20 million metric tons from their supply chain, one operational program they have introduced is the "Wal-Mart Supplier Greenhouse Gas Innovation Program." The program was introduced in collaboration with non-profits (Environmental Defense Fund and Carbon Disclosure Project), consulting firms (ClearCarbon and BluSkye), academic partners (Applied Sustainability Center at the University of Arkansas) and auditors (PWC) with a charge to identify products for potential GHG reductions, achieve reductions by engaging suppliers and customers and quantify and verify any GHG reduction claim.

The program identifies potential product categories by computing the "carbon intensity" of the product. Carbon intensity is simply the GHG emissions per unit multiplied by its sales. Categories with high carbon intensity are identified and projects for GHG reductions are initiated with the appropriate SVN Captains. The program team identified specific product(s), a Wal-Mart "Champion" (the lead project manager), upstream (supplier base), and downstream (customer base) activities to interact with. The Champion will lead the footprinting effort—map the supply chain, collect data, compute supply chain footprint, target areas for reduction, and clearly document actions and reductions in GHG emissions so they can be verified and audited.

Herman Miller's sustainability strategy has evolved around the Environmental Quality Action Team (EQAT), a cross-functional steering committee of Herman Miller employees that sets their environmental direction and priorities; and measures results.

The EQAT established several support teams to carry out specific tasks. A partial list includes: a Communications team (to better communicate their message), Design for the Environment (DfE) Teams (for setting environmental sensitive design standards), a ISO 14001 Team (focused on the continuous improvement of the Environmental Management System), and a Energy Reduction Team, committed to reducing the amount of energy necessary to produce Herman Miller products. According to Herman Miller, about 400 employees out of about 8,500 are involved directly or indirectly with EQAT initiatives.

When Herman Miller designed the Mirra chair, the DfE team was lead by the Director of Environmental Health and Safety with close partnership with a supply chain manager and a chemical engineer. The supply chain manager worked on supplier relationships while the chemical engineer provided input into the sustainability of the chemicals used[6].

Wal-Mart's Supplier Greenhouse Gas Innovation Program or Herman Miller's DfE teams are examples of operational execution of sustainability strategy, but for any footprinting exercise to be successful, the firm must (a) have an overall strategy

[6] The intent of this DfE team was to design the chair to "Cradle to Cradle" standards set by the McDonough Braungart Design Chemistry (MBDC) firm.

for sustainability and the responsibility must lie at the highest level—senior executive officers or the Board; (b) a process or program, like the Herman Miller's DfE team or Wal-Mart's Supplier Greenhouse Gas Innovation Program, by which such strategic objectives are executed and disseminated across the firm; and finally (c) operational details like the project manager, products or categories to be footprinted, timelines for GHG emission measurement, verification, audit, and its eventual mitigation need to be clearly defined prior to the start of the footprinting project.

12.2.2 *Mapping the Supply Chain, Data Gathering and Lifecycle Inventory*

Products or services to be footprinted are typically chosen based on the goals of the project. Cost savings, marketing opportunities, competitive considerations, ease of supplier engagement, and the potential time and effort needed are typically factored when considering products to footprint. Some companies (like Patagonia, Apple) footprint and publish GHG emissions for most products and facilities.

For the chosen product (service or facility), the project team needs to inventory the processes for which emissions are measured. Several standard accounting methodologies are now available such as ISO14040 and ISO14064 standards, WRI/WBCSD's GHG Protocol[7], and British Standards Institution's PAS 2050. The standards involve defining unit of product that is measured, setting boundaries on what gets measured, incorporating all appropriate supply chain activities, and accounting standards to allocate emissions to activities. This would necessarily involve engagement of supply chain partners and a framework by which footprints can be verified, analyzed, and action for reductions taken.

A typical lifecycle (Scope 1+2+3 emissions) includes the following (See Table 12.1):

Raw Material Extraction: The raw material product phase relates to the GHGs embodied within the raw material inputs that makes up a product, the emissions associated with their extraction, and transportation to a processing plant. For example, in the electronics and computer industry, extraction of metals for use in electronics and their shipment to component manufacturers would be a part of this phase. For the construction industry, logging and the eventual transport to lumber mills constitute the extraction phase.

Manufacturing: The manufacturing phase relates to the emissions released while transforming raw materials and other inputs into a finished product. In the electronics industry, this is the GHG emissions released in making of the components like the circuit boards, disk drives, memory chips, etc. and the assembly of these components into finished products. It is typical, at least for the electronics supply chains, for

[7] The WRI/WBCSD's GHG Protocol is currently developing standards to measure Scope 3 emissions.

Table 12.1 Sample life cycle activities sample activity breakdown

Life-cycle stage	Sample activities	Examples
Raw materials	Impact of raw materials	Deforestation, etc. Land use change
	Inputs used in the product	Bill of materials or teardown reports
	Typical processes	Extraction/mining Farming Livestock Logging Processing Transportation Waste streams
Manufacture	Inbound logistics	Processing Shipment of raw materials to Manufacture/assembly Storage
	Outbound logistics	Packaging Storage Transportation
	Production processes	Facilities Inputs and outputs of production Intermediate and final products produced Waste streams generated
Distribution/retail	Distribution operations	Processing Shipment to distribution center Storage
	Retail operations	Display Processing Shipment to retail/customer Storage/refrigeration
Customer use	Use	Maintenance Repair Use of product
End of life	Waste streams	Composting Land filling Recycling Take back/remanufacturing, etc.
Other considerations	Capital goods	Any capital to enable lifecycle activities
	Employees	Business travel Commuting
	Facilities	Franchises Leased and owned assets

multiple companies to be involved in this phase with multiple echelons in the supply chain so tracking the source of all the components can be challenging. Often firms in the supply chain use estimates—industry averages – to compute this phase of the life cycle.

Distribution: The distribution and retail phase details emissions from transporting finished goods to distribution centers and final retail locations. Typical emissions include distribution center operations, fuel combustion, truck and in-store refrigeration, and retail energy consumption.

Consumer Use: These are the GHG emissions from product use. These include electricity use for consumer electronics, fuel and emissions for the transportation sector, land and building use emissions for facilities, etc.

End-of-Life/New Life (recycle or other): The end-of-life product phase captures emissions resulting from waste and reuse streams like land fills, composting, remanufacturing, refurbishing and recycling.

12.2.3 Data Collection Challenges

There are typically two types of data collection. Activity data is simply the data on the type, number, and quantity of inputs and outputs; energy use; and net GHG emissions in each activity in the life cycle. Emissions data, on the other hand, are standard factors typically available in government or industry-specific databases that provide information to covert all activity data into the CO_2-eq. measure.

Activity data could be a result of direct measurement (for example, KwH used in production) or an industry estimate based of quantities of inputs/outputs used. The primary challenge for most reporting firms is to obtain direct activity data from suppliers and downstream partners. First, the reporting firms need to decide which suppliers to poll for activity data. Large firms have tens to hundreds of suppliers each with varying levels of activity data. It is typical to prioritize suppliers based on spend, the potential for GHG reduction, and the level of relationship. For smaller suppliers (relatively small spend), secondary data from well-established life cycle databases can be used to estimate emissions. Second, reporting firms need to ally any confidentiality concerns supply chain partners might have. Suppliers typically have multiple customers—so any activity or emission data that is requested is aggregated across their customer base. The supplier, using consistent standards, is best suited to allocate its emissions to the reporting firm—this may require confidentiality agreements to reduce concerns. Third, the reporting firm needs to provide a simple and practical template for data collection—a template that informs the suppliers what is needed and why; and the resources it would take to complete the footprint exercise. Finally, a designated contact at the supply chain partner is essential to clarify the data.

The goal of data collection (and indeed the entire footprinting exercise) is having accurate, consistent, and repeatable protocol so process improvement and eventual reductions in GHG emissions can be accurately measured and tracked.

Once the supply chain activities are mapped and activity and emission data collected, the footprint can be computed by allocating emissions to each activity in the supply chain and adding over activities in each stage of the life cycle. The various protocols—WRI/WBCSD's GHG Protocol and British Standards Institution's PAS 2050, for example, specify in great detail how the process map and related activity data can be used to produce the footprint.

During the data collection process, there are typically a lot of unresolved issues, for example, lack of activity data from partners or lack of industry standards (for example, average use of a phone or a washing machine). Assumptions or workarounds need to be documented and steps taken to clarify such issues in future considerations. It also helps to run a sensitivity analysis of the footprint—so the "average" is reported but depending on assumptions of material and product usage, the footprint can be more or less than the reported number.

12.2.4 Verify, Evaluate, and Act

Once the footprint analysis for a product or service is completed, it needs to be validated both internally within the reporting firm and externally by supply chain partners and third-party organizations. Protocols such as the GHG or the PAS 2050 provide guidelines on validating, auditing, and reporting the footprint. The goals of the footprinting process can be multi-fold. First, the reporting firm can use this as a baseline for carbon emissions. Any changes to product or process can be compared to this baseline to gauge the effectiveness of carbon mitigating projects. Second, the reporting firm can file the results in a public forum like their CSR reports or third-party programs such as the Carbon Disclosure Project to address the needs of their stakeholders. Third, they can use the footprint as tool to manage climate risks and act on carbon mitigation opportunities.

12.3 Using the Footprint as a Tool to Reduce Life Cycle Carbon Emissions

Footprints identify "hot spots" of carbon in a product's life cycle. Depending on the overall goals of the firm—risk mitigation, cost savings, brand image, etc.—changes can be made to the supply chain that reduces the footprint.

Since 2007, carbon footprinting has been promoted as an important tool for consumers to make sustainable consumption choices. However, sustainable development requires a fine global balance between the environment and development agendas of organizations. Food products are some of the first of "short life-cycle" products that organizations have used for consumer facing carbon footprinting efforts. For many developing countries, complete carbon emission based food pricing would be a desirable option wherein crops that are grown under the sun with limited chemical inputs tend to have lower emissions. But methodologies such as LCA that are used tend to favor easy measurement and examine only product-based emissions without involving systemic emissions.

The footprint gives a way to meaningfully engage supply chain partners to mitigate climate change issues. Depending on where in the supply chain the best returns in carbon mitigation investment are, firms can prioritize whom they engage and what the engagement should entail.

Table 12.2 Sample footprint reduction strategies

Category	Action steps
Material inputs	Use fewer variety Use less quantity Eliminate hazardous substances Use materials that can be reclaimed Use materials that can be "upcycled" Sustainable extraction (logging, seafood, mining, etc.)
Energy use	Move to renewable energy Reduce energy consumption Move to efficient fuels (Electricity to Gas)
Production process	Locate production close to supply Use fewer processing steps Increase efficiency in process Reduce waste streams (solid, water, etc.)
Distribution/retail	Reduce distribution distances Fuel efficient transportation, backhauls Energy efficient distribution/retail refrigeration Low GHG retail design
Facilities/employees	Energy and water efficiency (e.g., LEED certification) Employee commute programs Expansion of green spaces Active involvement from employees
Consumer use/end of life	Design for energy efficient use Establish product recovery steams at end of life
Supply chain relationships	Select suppliers based on GHG criteria Educate suppliers on GHG reduction benefits Joint programs for carbon reduction Invest in supply chain partner process to mitigate carbon Educate consumers, drive low-GHG product choice

Common strategies are illustrated in Table 12.2. One class of strategies involves changes to material (raw material) inputs into processes. Choosing sustainable inputs that are safe and reclaimable, responsibly extracting it, and its parsimonious and efficient use are all finding their way into supply chain strategies. Wal-Mart, for example, is working with its seafood suppliers to have its Atlantic salmon Marine Stewardship Council (MSC)-certified, a protocol that ensures responsible fishing and farming practices[8]. Nokia's N8 Smartphone[9] uses bio plastics and is free of toxic substances like PVC, BFR, and RFR, making all the materials recyclable. Nokia requires its primary suppliers to clearly state the chemistry of components it sources. Although Apple's current 21.5 in. iMac is more powerful and has a much larger screen than the first-generation 15 in. iMac, it's innovative design uses 50% less

[8] http://walmartstores.com/Sustainability/10607.aspx?p=9173.

[9] http://www.nokia.com/environment/devices-and-services.

material and generates 35% fewer carbon emissions[10]. Shaw's EverTouch® nylon flooring products are made from Type 6 nylon that can be recycled back into carpet fiber again and again, endlessly (also called "upcycling"), closing the loop in the material supply chain[11].

A second major direction in reducing footprint is to move towards renewable energy (or reduce energy reliance on fossil fuels), simply reduce the use of energy and eliminate any waste (including water and direct GHG emissions) from supply chain processes. This involves not only the reporting firm's processes but also relevant activities in the supply partners' processes. Such efficiency-related projects can provide quick short-term gains to the firm. For example, as of 1 May 2010, Herman Miller uses 100% renewable electrical energy in all their facilities worldwide[12]. They also reduced their electricity consumption from 100 million KwH in 2007–2008-85 million KwH in 2009–2010. In 2010, Wal-Mart's China operations, for example, reduced plastic shopping bag waste in their stores by 84% when compared to a 2007 baseline. Through water conservation programs they have reduced water usage by 54.2% since 2005[13]. Google uses efficient servers and innovative cooling methods in their data centers to save 50% energy over comparable data centers[14].

A third thrust in reducing GHG emissions is making facilities energy efficient and employees engaged in the carbon mitigation strategy. Several third party certification systems exist for green buildings. The most recognized among them is LEED, developed by the U.S. Green Building Council (USGBC). According to the USGBC, LEED provides "building owners and operators with a framework for identifying and implementing practical and measurable green building design, construction, operations and maintenance solutions." Examples include Herman Miller and Adobe's corporate headquarters, Patagonia's Reno distribution center, and the World Financial Center in Beijing, China.

Hand-in-hand with facilities design, firms are also increasingly getting employees engaged in carbon mitigation activities. Google has bio-diesel commuter buses to transport employees to work. Within the Google campus, Google has a program called "GFleet"—a fleet of plug-in cars to shuttle employees across their campus. According to Google these programs have saved 5,400 tons of CO_2 or an equivalent of 14 million cars off the road[15]. Programs at other companies include providing incentives for public transportation, carpooling, committing to a green lifestyle (for example, Saatchi and Saatchi's "Do One Thing" (DOT)). The goals are not only carbon mitigation, but also to boost employee morale to increase productivity.

Finally, carbon mitigation programs also involve consumer education and engagement to help save energy, reduce direct emissions, and close the material loop in the supply chain. Patagonia's "Common Thread Initiative" encourages customers not only to reduce, repair, reuse their products to extend the life of the product (they

[10] http://www.apple.com/environment/.

[11] http://www.shawfloors.com/Environmental/EnvironmentallyFriendlyDetail.

[12] http://www.hermanmiller.com/About-Us/Environmental-Advocacy.

[13] http://www.apple.com/environment/.

[14] http://www.google.com/green/operations/data-center.html.

[15] http://www.google.com/green/operations/transportation.html.

back this up by making high-quality products with sustainable inputs), but provides a network of retail outlets through which customers can recycle their garments. Electronic manufacturers like Dell, HP, Apple all design energy efficient products and provide guidelines on how to use the product in an energy-efficient way. At the end of life, they all take back their products so they can be recycled and/or disposed in an environmentally safe way.

12.4 Examples of Complete Life Cycle Product Footprints and Actions

12.4.1 NIKA: Bottled Water

NIKA's published mission is to fight global poverty by selling premium bottled water. They donate 100% of their profits to support clean water projects in under-developed countries. According to NIKA they "will provide the basic tools and critical assistance to help thousands of families improve their lives in a meaningful way and end the cycle of poverty."

NIKA has undergone a rigorous third-party audit of their manufacturing and distribution process to determine their GHG emissions. What they discovered were the following per-bottle CO_2-eq. emissions[16]:

- Raw materials—73 g
- Materials manufacturing and transport—43 g
- Product manufacturing and transport—105 g
- Consumer use—2 g
- End of life—3 g
- Total—226 g

Since product manufacturing and transport are the largest part of the footprint, NIKA is lowering their footprint by localizing their bottled water production facilities. By sourcing, purifying, and bottling their water in a handful of geographically differentiated locations throughout North America, they are able to minimize the use of transportation.

They also have a "One-for-One" recycling program, where they pledge to recycle one plastic bottle for each one they sell. They also pay schools a fee (to be donated to charities) to recycle plastic bottles.

Working with Carbonfund.org, NIKA now offsets their product's carbon footprint by contributing to a reforestation project in Nicaragua. As a result, they have achieved "CarbonFree" certification from Carbonfund.org, allowing them to claim that they are a carbon neutral company—the first of its kind in the bottle water industry in the US.

---

[16] http://www.nikawater.com.

12.4.2 Pepsi Co.: Tropicana Orange Juice and Walkers Crisps

Pepsi Co., working with the Columbia Earth Institute computed footprint of a half a gallon carton of Tropicana orange juice as 3.75 lb.[17] of CO_2-eq.

By far, the largest GHG contribution actually came from the cultivation of orange trees. According to Pepsi Co., 60% or 2.25 lb. of the carbon footprint of a half gallon of Tropicana orange juice are produced during the fruit growth and juice production stage.

The production-related emissions include the production and application of the nitrogen fertilizer, picking and transporting the oranges to the Bradenton, FL processing plant, as well as the use of energy to squeeze, pasteurize, and pack the juice.

The 64 oz. cartons are shipped by train throughout the country—22% of the GHG emissions are produced during the distribution phase. And finally, they calculate the packaging at 15% and use and disposal at 3%.

Since 35% of the footprint is from nitrogen fertilizer applied to orange groves, Pepsi Co., in collaboration with one of its growers, SMR Farms in Bradenton, Fla., has launched pilot studies to test two low-carbon fertilizers to determine whether using either could significantly reduce the carbon footprint of growing oranges. If successful, Pepsi Co. projects that this change could reduce the total carbon footprint of Tropicana Pure Premium by as much as 15%.

Walkers Crisps Walkers[18] is another of Pepsi's biggest brands, and a best selling item is the cheese and onion potato chips (crisps in the UK).

Pepsi Co. examined every stage in the production of these crisps, beginning with growing potatoes, sunflowers (for oil), and seasoning, which made up 34% of the footprint. Production of the chips was responsible for 17% and packaging accounted for 36% of the total footprint.

Transportation of the crisps to grocery shelves comprised 10% of the footprint. The remaining 3% goes to disposing of the empty packaging from the crisps. All in all, they estimate that one bag of crisps results in the emission of about 80 g (about 0.18 lb.) of CO_2-eq. emissions.

Since a third of the greenhouse gas emissions are during the production cycle of the supply chain, PepsiCo has been working with its suppliers to develop methods of reducing energy consumption in this phase of the project such as capturing water extracted from potatoes to achieve zero water intake, using renewable energy at some sites, and reducing waste sent to landfills by 39%. The footprint also helped redesign some procurement practices. Since Walkers was buying its potatoes by gross weight, farmers were storing their potatoes in humidity controlled facilities to increase water content. This necessitated Walkers to fry the sliced potatoes longer to remove the extra moisture. Buying potatoes by dry weight provided a win–win situation—Walkers could reduce frying time and farmers could avoid the cost of

[17] http://www.nytimes.com/2009/01/22/business/22pepsi.html?ref=business.

[18] http://www.walkerscarbonfootprint.co.uk/walkers_carbon_footprint.html.

humidification. All these process changes resulted in a reduction of greenhouse gas emissions by about 7% between 2007 and 2009, saving the company £400,000.

12.4.3 Apple: iPhone4

Apple reports the iPhone4's greenhouse gas emissions[19] as follows:

- Production—57%
- Customer use—34%
- Transport—8%
- Recycling—1%

Apple has suppliers scattered throughout the world[20]. For example, the iPhone4's display (LG), processor and memory (Samsung) are sourced from Korea. The electronics come, among other countries, from Japan, Germany, Taiwan, and the US. The phone is assembled in China and shipped via air worldwide. The sourcing, manufacture, and assembly accounts for 57% of the 99 lb. CO_2-eq. footprint. Transportation for final sale to the customer contributes 8%. Customer use accounts for 34%[21] and recycling at end of life 3% to the footprint.

Since the energy consumed during product use is a major source of GHG emissions, Apple has designed the iPhone4 with power conservation in mind[22]. For instance, the power adapter is ENERGY STAR qualified, and the components and power management software of the iPhone4 are configured to use energy as minimally as possible.

The compact design uses less material inputs, and the packaging is minimized and is almost 100% recyclable. The iPhone4 also makes disposal and reclamation of raw materials easier by designing the phone free of toxic materials such as arsenic, BFR, mercury, and PVC.

Apple also has a comprehensive take back program—customers can mail in their product to Apple and Apple will recycle it in a responsible way.

12.4.4 Patagonia's Nano Puff Pullover Jacket

Patagonia tracks the footprint of its products and publishes them as "Footprint Chronicles" on their website. Their Nano Puff Pullover Jacket[23], for instance is designed in

---

[19] http://www.apple.com/environment/reports/.

[20] http://operationsbuzz.com/2010/11/the-iphone-4-supply-chain/.

[21] Apple assumes a use phase duration of 3 years for the calculation of handheld power consumption. Since there is a lack of a universal benchmark, Apple estimates of the amount of time and the intensity of power drawn during the use phase.

[22] The previous model iPhone3GS had a footprint of £ 121.

[23] http://www.patagonia.com/web/us/footprint/index.jsp.

Ventura, CA. Its design objective was to provide the most warmth to weight ratio, yet make it sustainably. Fibers and fabric for this jacket are sourced in Nobeoka, Japan and the "Primaloft," a polyester-based lightweight insulation, is made in Rudong, China. The garment is sewn in Ho Chi Min City in Vietnam and finally distributed to the retailers via their distribution center in Reno, NV.

Patagonia estimates that it consumes about 43 MJ to make the jacket, which is equivalent to burning an 18 W light bulb 24 h per day for 28 days.

The reported footprint for this garment is about 5.5 lb. Patagonia has included much of the upstream activities covered by GHG Protocol including pick-up of the original fibers to manufacturing power consumption through its 12,545 mile journey to the distribution center in the US. This footprint does not include downstream GHG emissions of retail operations, customer use, and recycling and retrieval of the garment at the end of its life.

The Nano Puff uses 63 L of water and the waste stream is 2.7 oz, about the third the weight of the Jacket. According to Patagonia, transportation was only a small portion of the footprint—so as long as the jackets are transported by ship, making them in Ho Chi Min City provided the economic benefit of lower labor cost. The largest portion of the footprint was in the manufacturing process. The company is working with their suppliers and their manufacturing partners to improve processes.

The jacket is fully recyclable at the end of its life. Patagonia can reclaim the shell and lining (which are made from recycled content) and convert it back to virgin raw material. The PrimaLoft insulation as part of the Common Threads Initiative is completely recyclable; but virgin polyester is used to make the insulation in the jacket. Patagonia is working with Albany International, the maker of PrimaLoft, to develop lightweight and compressible insulation from recycled materials. A coating on the jacket (Patagonia calls it DWR) to repel water produces Perfluorooctanoic acid (PFOA), which is considered toxic. Patagonia is also working with suppliers to eliminate PFOA in the production process.

12.4.5 Toyota Pruis

The best selling Toyota Prius has a footprint of 44 metric tones or 97,000 lb.[24] over its 120,000 mile life. The footprint includes sourcing of parts, car assembly, customer use, and the eventual disposal of the car. Upstream supplier operations and assembly accounts for 18.6% of the footprint. The largest part of the footprint is making of the fuel and its use in the car, which accounts for 68.5% of the footprint.

In this case, fuel efficiency of the car primarily determines the footprint. The fuel efficiency of the Prius is estimated at 42 miles per gallon. In comparison, the Toyota Camry with an efficiency of 29 miles per gallon has a footprint of 64 tons.

---

[24] http://online.wsj.com/public/resources/documents/FOOTPRINT.pdf.

The 4 Runner, an SUV from Toyota has a fuel efficiency of 16 miles per gallon and a footprint of 118 tons[25].

12.5 Concluding Remarks on Working with Supply Chain Partners

Working with supply chain partners on carbon mitigation projects can be challenging. First, suppliers may not track activity data or their GHG emissions and as a result may not have reduction targets in their strategic plans. Large reporting firms, however, can use GHG criteria to select suppliers, in effect forcing many of the suppliers to comply[26]. Dell expects its primary suppliers to track and manage GHG emissions. In fact, Dell insists that their key suppliers report their emissions in public third party programs like the Carbon Disclosure Project. Dell also expects its suppliers to set specific goals not only for GHG reductions in its own operations but enforce GHG reduction on its suppliers (Dell's Tier 2 suppliers). Lack of compliance could entail a lowered quarterly business review score, crippling their ability to compete for Dell's business.

Proctor & Gamble (P & G) has a supplier scorecard that tallies the performance of suppliers on multiple sustainability criteria like the use of energy and water and the management of material and waste, and rewards suppliers with higher ratings if they collaborate with P & G initiatives on climate-change issues. An added benefit of this scoring system is that the suppliers not only align their GHG emission strategies with P & G but also use the scorecard to assess their own suppliers, as it directly impacts their rating with P & G.

Wal-Mart, is developing a Sustainability Index that asks 15 questions on various areas of sustainability to its 60,000 + suppliers. Wal-Mart wants the suppliers to report their results directly through the CDP Supply Chain Program. The sustainability index will be used as tool to score suppliers and products as well as identify opportunities for future collaboration.

For smaller firms who cannot use their size as leverage—if larger suppliers find it uneconomical to change their process for its smaller volumes—the solution is to redesign products with a different set of more sustainable materials that are readily available. For example, Nau, a small apparel manufacturer based in Portland, OR, uses wool, recycled polyester, and organic cotton as inputs to their garments to make them sustainable. They set standards for suppliers to follow, but to a large extent they use off-the-shelf fabric and existing supplier processes to manufacture garments. They engage their customers by donating 2% of every purchase to sustainable causes.

---

[25] The Camry and 4Runner footprints were reported in http://online.wsj.com/article/SB122304950601802565.html.

[26] According to the 2011 Carbon Disclosure Project's Supply Chain Program, 12% of respondents indicated that they are deselecting suppliers based on GHG criteria. More importantly, 56% of them reported that they would use sustainability criteria to select suppliers in the next five years.

By far, the most prevalent model[27] is one of collaboration where the supply chain partners jointly collaborate on projects to reduce GHG emissions. The reporting or the lead firm initiates the project, opportunities are recognized, and a joint program is developed with agreed upon targets (both qualitatively and quantitatively). For example, in 2005 Wal-Mart initiated a project with Unilever to design and market "all® small-and-mighty," in 32 oz. containers which is three-times concentrated, and lasts the same 32 loads as a 100 oz. bottle[28]. The success of this initiative led Wal-Mart to work with other detergent suppliers like P&G, Dial, Huish and Church and Dwight, to transform their production processes to produce concentrated detergent. According to Wal-Mart this initiative saved 430 million gallons of water, and eliminated the use of 80 million lb. of plastic resin and 125 million lb. of cardboard.

Such examples of collaboration are numerous—supply chain partners need to align their sustainability-related incentives to make the project a success. P & G spent almost $ 200 million to retool its processes to make concentrated liquid detergent to meet Wal-Mart's goals—but anticipates that savings in materials, energy, water, and waste reduction will pay for such investments over time.

Often, reporting firms work and invest in their supplier's processes to transform them. Herman Miller's DfE team had to work with their suppliers to identify sustainable material for the Mirra chair. The furniture industry had well-established supply chains for materials (for example, plastic—its manufacture and molding into shapes) and any change will necessitate suppliers to invest in technology and material development. Herman Miller worked with over 200 suppliers to help identify sustainable materials and possible substitutes—a significant investment in time and resources.

Firms also face a number of operational challenges when working with partners. First, they have to agree on data and reporting standards. Since multiple standards are in use around the world, they need to reconcile GHG emission assumptions. Second, firms also need to prioritize how to set reduction targets. For example, targets can be low but can involve more partners in the supply chain; or they can aim for larger reductions working with select suppliers in a narrow slice of the supply chain. Third, firms also need to agree on how to report emissions, reduction targets, mitigation actions, and how if any, share the benefits of joint sustainability programs. While there are several industry-wide or third party based initiatives to resolve such issues, standards and protocols that tackle the measurement of GHG emissions, mitigation, and reporting from a supply chain perspective are still in their infancy.

However, as stakeholders demand life cycle emission information, and as firms discover the benefits of sustainable production by reducing their footprint, we expect the supply chain or the life cycle approach to gain wide acceptance in the near future.

[27] 46% of CDP Supply Chain Program respondents have collaborative projects. In five years this number is expected to rise to 86%.

[28] http://walmartstores.com/media/factsheets/fs_2328.pdf.

Chapter 13
Making the Business Case for Sustainability Related Investments Through a Single Financial Metric

Mark Ferguson

Abstract Most firms have set procedures or thresholds for evaluating capital-intensive projects, such as a minimum level for the payback period or ROI on the project investment. While there is often uncertainty involved in projecting the cost savings or increases in revenue associated with the proposed project, for the most part, reasonable estimates exist for the future unknown parameter values that are needed to make the business case. In today's business environment, almost every firm is starting to include sustainability-related projects in their regular review cycles. Investments in sustainability-related projects often include some type of cost savings associated with a decrease in the use of a non-renewable commodity such as oil, ore, or water. In these situations, managers often struggle when presenting their business case, in a manner that is understood and consistent with other investment options. One aspect that makes it particularly difficult is fuel, energy, and raw materials are commodities, whose prices fluctuate considerably; thus the resulting yearly savings from the more sustainable technology depend on the underlying assumptions about these prices. In this chapter, we show how the price of options to purchase the commodity at various stages in the future at today's cost can be used to place a financial value on the savings attributed to sustainability-related investment. This technique provides a single, bottom-line number that managers are more familiar with when evaluating where to allocate a fixed investment budget among competing project proposals.

13.1 Introduction

There is little debate today that environmental issues will play a major role in companies' long term prospects. While environmental issues, such as carbon dioxide emissions and climate change have received the vast majority of the attention in the popular press, a less publicized, but just as critical, issue is the increasing rate of sending manufactured products made of non-renewable raw materials and resources to landfills and incinerators around the world. While the practice of a "disposable

M. Ferguson (✉)
Moore School of Business, University of South Carolina,
1705 College Street, Columbia, SC 29208, USA
e-mail: Mark.ferguson@moore.sc.edu

T. Boone et al. (eds.), *Sustainable Supply Chains*,
International Series in Operations Research & Management Science 174,
DOI 10.1007/978-1-4419-6105-1_13, © Springer Science+Business Media New York 2012

society" is common today, it has not always been so. Prior to the early 1900s, very few consumer goods were sent to landfills or incinerated. Wood and metal from a wagon that had reached the end of its useful life, for example, would be salvaged to build houses or make furniture. The first modern landfill in the US was not created until 1937. Prior to this time, the small amount of household waste was simply composted or burned.

Consumer habits toward waste started to change, however, in 1901, the year King Camp Gillette started his disposable razor company. King Gillette's innovative idea was to create an inexpensive replaceable blade for his razors that could simply be disposed of when it became dull. Thus began the era of the disposable society. Today, advances in manufacturing and supply chain management have brought the cost down on most consumer goods to the point it has become cheaper to dispose and replace than to repair and reuse. Even with technologically sophisticated items such as DVD players, computers, cellular phones, or microwave ovens, the technology changes so quickly there is little value in expending effort to extend their useful life for more than a few years from the date of manufacture. The consequence of this change in behavior is an exponential increase in the demand for non-renewable resources around the world, to the point that many are arguing we are reaching the tipping points, where the supply of non-renewable raw materials is unable to keep up with the increasing demand (for example, peak-oil).

A disposable society cannot continue forever, however. Eventually, the amount of non-renewable raw materials and the land available for product disposal will be used up, leading to a significant reduction in the fortunes of pure product-based companies and a lower standard of living for consumers around the world. The numbers illustrating the problem are hard to fathom. Each household in the UK generates approximately 1 ton of waste each year. Even worse, for every ton of products we buy, 10 tons of resources are used to produce them[1]. In the US, each person generates approximately 4.5 pounds of waste per day for a cumulative total of 250 million tons of solid waste that were either incinerated or sent to landfills in the year 2008. Of these 250 tons, 18% were categorized as durable goods. The disposal of durable goods is particularly troublesome because they are often manufactured using material from non-renewable resources. Unfortunately, of the 46 million tons by weight of durable goods sold in the US in 2008, only 17.4% of the material used in their production has been, or is expected to be, recovered. As a consequence, landfills are filling up around the world. Incineration of waste has its own problems, as the process produces toxic pollution and emits gasses that contribute to the global warming problem. Finally, our finite supply of virgin raw materials are being extracted from the earth and chemically processed into alternative forms to make products that are quickly disposed of and then sent to be burned or mixed with other toxic materials in landfills. Simply put, the current business practice of extracting raw materials from the earth, manufacturing them into products, and then disposing of the products into landfills or incinerators after a short period of use is not a sustainable practice.

Some enlightened firms recognize this fact and are starting to devise strategies for their long-term survival; strategies that involve dramatic changes in the way they

[1] http://www.wasteonline.org.uk/resources/InformationSheets/HistoryofWaste.htm

have historically viewed their supply chains. A promising trend in recent years has been the number of firms who make investments in non-renewable resource saving technologies and who pro-actively provide recycling options for their products at the end of the product's useful life. Many of these initial investments have resulted in changes that improve both the firm's environmental footprint as well as the bottom line. Often times, it is easy to justify these initial projects because the total investment is relatively small (for example changing over to more efficient lighting) or the financial and environmental improvements are evident, even under a worse-case scenario (for example non capital intensive energy efficiency improvements). These initial projects often represent the "low hanging fruit" of sustainability-related investments, however, and firms quickly run into more challenging sustainability-related investment options where the environmental benefits are often clear but the financial benefits are not. Often, these projects will result in some type of energy or raw materials savings over the life-span of the asset but the total financial benefit depends on an uncertain future cost of the energy or raw material the project is projected to reduce the usage of. Under a pessimistic (typically low) cost estimate; the project may have a negative ROI while under an optimistic (typically high) cost estimate; the project has a positive ROI. Thus, the decision to invest in this project is not clear cut. These are the type of projects that we will focus on in this chapter.

While the project sponsors can present the case for the investment using a triple-bottom-line approach (showing the benefits to people, profit, and the environment), most firms have a limited budget for investing in their business and the sustainability-related project will be competing against alternative investment opportunities whose benefit is measured solely on the projected financial impact. In such cases, the sustainability-related project needs to have a financial impact that at least makes it competitive with the alternative investment options before the other benefits (people and environmental) are considered. Thus, the financial business case is still often the most important component that determines if the project receives funding or not.

We have observed that managers struggle to present a business case to the senior management team for such investments, in a manner that is understood and consistent with other investment options. One aspect of this scenario that makes it particularly difficult is fuel, energy, and raw materials are commodities, whose prices fluctuate considerably; thus the resulting yearly savings from the more sustainable technology depends on the underlying assumptions about these prices. In the next section, we discuss how the financial business case for these investment opportunities is most often made in practice (based on our own observations). We then present an alternative methodology which presents the business case in terms of a single financial number or metric, which is more consistent with the way the alternative investment options will be presented.

13.2 Current Methods for Making the Business Cases

We will present the typical business case methodology through a specific example, although similar examples can be constructed for any sustainability-related investment opportunity that requires a higher initial investment but promises future savings

Table 13.1 Example of possible analysis for the purchase of a hybrid delivery truck over a traditional diesel truck

	Average Fuel Cost Per Gallon Over 10 Year Expected Lifetime		
	−10% Current Price	Current Price	+10% Current Price
Payback period (years)	7.5	6	4.5
ROI	−3%	4%	12%

in some non-renewable resource. A carbonated beverage company is considering replacing their fleet of older delivery trucks with either new hybrid delivery trucks of the same capacity or new traditional delivery trucks. Traditional delivery trucks cost approximately $ 60,000 each, whereas a hybrid truck costs $ 90,000, a 50% increase in the initial investment. Depending on the driving conditions, a hybrid truck may provide significant savings in fuel over its life span as compared to a traditional delivery truck. The reader is referred to Hoyt and Plambeck (2006) for an overview of the hybrid truck industry and a description of the how the Eaton Corporation developed a hybrid delivery truck in response to a bid opportunity for Federal Express. Hybrid trucks provide the most benefit for medium range deliveries. Hybrid technology recovers energy from the vehicle's braking system, so deliveries over longer distances, which often involve travel over interstate highways generally won't benefit from the technology, while very short delivery routes are often in congested urban areas where the truck is sitting idle in heavy traffic. A typical medium range route for the firm's truck covers approximately 18,000 miles per year. Thus, the fuel savings of a hybrid truck over a traditional truck is approximately 100 gallons of diesel fuel a month.

Quantifying this savings of 100 gallons of diesel fuel per month is the main challenge in preparing a business case for this type of investment opportunity. A current common practice is to assume optimistic, current, and pessimistic scenarios for fuel prices over the truck's life span, and provide a payback period and ROI for the investment under these three scenarios. Based on the calculations summarized in Table 13.1, the payback periods range between 4.5 and 7.5 years, depending on fuel price assumptions. The corresponding ROIs range from 12% to –3%. Suppose the firm has typical thresholds to determine if projects are funded for the payback period of 5 years and an ROI of 10%. In this scenario, determining if the project is funded or not depends on which estimate of future fuel prices the management team has the most confidence in—a difficult task for most management teams who are not typically experts at predicting future commodity prices.

This simplistic approach to presenting a business case has a couple of disadvantages: (i) it does not consider the underlying uncertainty of fuel prices from year to year, and (ii) it does not provide a single payback number that other alternative investments have, thus putting it at a possible competitive disadvantage for selection and funding among competing alternatives. As an example, suppose the hybrid truck project is competing against another (non-sustainability related) project (replacing an existing delivery truck with a new non-hybrid truck, for example). Since fuel savings is not a significant reason for replacing an end-of-life delivery truck with a non-hybrid truck, the non-hybrid truck has a less uncertain payback period of 4.8 years and an ROI of 11% (based on factors such as reduced maintenance cost and a reduction in downtime). Since most management teams prefer less uncertainty (or at least the illusion of it), they may select the alternative project over the hybrid truck project, since the hybrid truck project only meets the thresholds at the high fuel cost estimates.

Another common approach that does provide a single number is to assign probabilities to each of the three future fuel cost scenarios. Using the example from Table 13.1, the firm could assign a probability of 0.2 for the case of the average fuel cost decreasing by 10%, 0.2 for the case where average fuel prices stay the same as today's price, and 0.6 for the probability that the average fuel cost will be 10% higher than today's cost. While this approach does provide a single number for the payback period and a single number for the ROI, there is often little justification for how these probabilities are assigned. Thus, this approach can be even less informative than the alternative scenario approach shown in Table 13.1. In the next section, we show an alternative way to present the business case for the hybrid truck that uses the experts' projections of future fuel costs and provides a single payback period or ROI metric.

13.3 Proposed Method for Making the Business Case

We propose a methodology that incorporates the prices of existing call options markets for fuel to quantify the savings into today's dollars. Our proposed approach addresses the two main drawbacks of the traditional approaches mentioned above, the lack of expertise in projecting future commodity prices and the lack of a single "bottom-line" number. It can be applied to any investment where there is an environmental alternative that saves on raw materials, fuel, and energy which are traded on existing futures markets. In situations where call options are not currently traded for the commodity, the approach can still be used by manually calculating the value of call options using the methodology described in Lemoine (2009), who values the fuel flexibility of plug-in hybrid electric vehicles by representing the purchase of the battery as the purchase of a strip of call options on the price of transportation. To value the call options, he uses a Kalmanfilter forecasting method to find the maximum likelihood estimates for three gasoline price models applied to a US municipal market.

For the hybrid truck example above, suppose that an initial evaluation showed the metrics for replacement of the current delivery trucks with new delivery trucks meet the firm's minimum threshold levels for investment. Thus, the "do not invest" option is eliminated from consideration and the remaining decision is whether to invest in new regular delivery trucks or their hybrid equivalents. We will simplify the decision to demonstrate our approach by assuming that the only difference between the regular delivery trucks and the hybrid trucks is the higher upfront cost for the hybrids and the fuel usage over the 10 year expected lifetime of the trucks. While there is no uncertainty around the higher upfront cost, the savings in dollars from the monthly fuel savings depends on the value chosen for the expected fuel cost, as described in the last section. As previously discussed, the uncertainty in future fuel prices makes this value difficult to quantify. This is where our new approach comes in.

To quantify the uncertainty involved in future fuel cost, the firm can purchase call options for 100 gallons of diesel for each month (the average savings of the hybrid truck over the traditional truck) over the truck's ten-year life span. The current value of the sum of these options should be added to the upfront investment of the non-hybrid truck so an easier financial evaluation can be made. To demonstrate, suppose the proposed purchase of the delivery trucks will take place in the current month where the current price of diesel fuel is $ 3 per gallon. To evaluate the financial business case of purchasing the hybrid truck over the regular delivery truck, the firm should sum up the price of call options for purchasing 100 gallons of diesel fuel at $ 3 per gallon for the current month plus one month all the way to the current month plus 120 months, as demonstrated by the following formula.

$$\textit{Total fuel savings} = \sum_{i=1}^{120} (\textit{price of call option for purchasing } 100 \textit{ gallons at } \$\,3 \textit{ in month } i)$$

The firm does not actually have to purchase these call options, they only need to know what their prices are. The sum of these call options should then be added to the upfront cost of the regular (non-hybrid) delivery truck when comparing against the upfront cost of the hybrid truck. Suppose that the total fuel savings over the 10 year expected lifetime of the delivery truck is $ 25,000 in today's dollars. This number can then be used to offset the $ 30,000 premium that the hybrid delivery truck commands (after deducting for any tax incentives) over the initial cost of the hybrid delivery truck. Thus, from a purely financial standpoint, the traditional delivery truck is still superior but the difference can now be stated as a single number, in this case $ 5,000. This difference is likely small enough that the management team will give the hybrid delivery vehicle additional consideration and potentially decide that the more intangible benefits (such as better public relations, improved employee morale, etc.) can tip the decision toward purchasing the hybrid vehicles.

13.4 Conclusion

Investments for projects that are targeted to reduce the use of non-renewable resources are often difficult to build a financial business case for because of the uncertainty in the future cost of the resource that will be conserved. Traditional approaches to building the business case involve providing multiple numbers of the same metric based on different assumptions about the resources' costs. The main drawback to the traditional approach is the multiple numbers for the metric have to be interpreted—and used to make a final investment decision—by a management team that seldom has expertise in projecting future costs of the resource the project is designed to conserve. Thus, sustainability related projects often face a disadvantage in the budget allocation process compared to more traditional projects where the benefits can be quantified more easily, and with less uncertainty.

Our proposed approach addresses this problem by using the price of call options (established by experts on the pricing of a particular non-renewable resource) to quantify the financial benefits of non-renewable resource conserving project. The approach results in a single number for a performance metric, such as payback period or ROI, which is more consistent with the traditional projects that the sustainability related project will be competing against. While our focus in this chapter has been restricted to the financial benefit of a sustainability related project, we acknowledge there are additional, non-financial, benefits that should also be included in the final budgeting decision. Without a credible financial analysis to compare the proposed projects, however, we fear many attractive sustainability-related projects do not make the initial financial cut-off and therefore do not receive further consideration by the management team. It is our hope that this new approach for quantifying the financial benefits will lead to more thorough evaluation from management of these important projects.

References

Hoyt D, Plambeck E (2006) Fedex and environmental defense: building a hybrid delivery fleet, case writing office at the Stanford Graduate School of Business

Lemoine D (2009) Valuing plug-in hybrid electric vehicles' battery capacity using a real options framework, University of California at Berkeley Working Paper

Chapter 14
Incorporating Life-Cycle Economic and Environmental Factors in Managerial Decision-Making

Ravi Subramanian

Abstract Recent environmental trends, including (1) an expansion of existing command and control directives, (2) the introduction of market-based policy instruments, and (3) the adoption of extended producer responsibility, have created a need for new tools to help managerial decision-making. To address this need, we develop a nonlinear mathematical programming model from a profit-maximizing firm's perspective, which can be tailored as a decision-support tool for firms facing environmental goals and constraints. We typify our approach using the specific context of diesel engine manufacturing and remanufacturing. The approach allows the incorporation of traditional operations planning considerations—in particular, capacity, production, and inventory—together with environmental considerations that range from product design through production to product end of life. A current hurdle to implementing such a model is the availability of input data. We therefore highlight the need to involve all departments within businesses and for industrial ecologists and business managers to work together to implement meaningful decision models that are based on accurate and timely data and can have positive economic and environmental impact.

14.1 Introduction

In their quest to tackle environmental impacts at various stages in the production and distribution of manufactured goods, regulators continue to implement a spectrum of environmental policies. As a result, companies face complex tradeoffs in dealing with policy and competitive mosaics. For example, manufacturers of electrical and electronic equipment in Europe now have to comply with the Waste Electrical and Electronic Equipment (WEEE, Directive 2003/108/EC) and Restriction of Hazardous Substances (RoHS, Directive 2002/95/EC) directives, in addition to the myriad stipulations on the release of waste during production into discharge streams (European Commission (EC) 2011). The WEEE directive aims to encourage reuse,

R. Subramanian (✉)
College of Management, Georgia Institute of Technology, Atlanta, GA 30308, USA
e-mail: ravi.subramanian@mgt.gatech.edu

T. Boone et al. (eds.), *Sustainable Supply Chains*, International Series in Operations Research & Management Science 174,
DOI 10.1007/978-1-4419-6105-1_14, © Springer Science+Business Media New York 2012

recycling, and recovery, and to improve the environmental performance of all actors involved in the life-cycles of electrical and electronic equipment. The RoHS directive restricts the use of lead, mercury, cadmium, hexavalent chromium, and polybrominated biphenyls or diphenyl ethers in electrical and electronic equipment. Regulators have also implemented industry-specific legislation such as the EC's End of Life Vehicle directive (ELV, Directive 2000/53/EC) that requires automobile manufacturers to take back vehicles at the end of their useful lives. The primary rationale for such legislation is to provide incentives to manufacturers to design environmentally more benign products. In the US, several states including California, Maine, Maryland, and Washington, have implemented various environmental laws providing similar incentives to manufacturers.

Within the past few years there has been a trend to supplement existing and planned command-and-control rules with market-based programs, which further complicates but also expands the array of choices that managers have in responding to environmental pressures. The first major market-based program was the cap-and-trade system for sulfur dioxide and nitrogen oxides under the US Clean Air Act Amendments of 1990 (US Environmental Protection Agency (EPA) 2005). Although this legislation affects only the US electric power generating industry, its success has motivated the adoption of cap and trade programs for a number of other pollutants around the world. The most notable example to date is the greenhouse gas Emissions Trading Scheme (ETS, Directive 2003/87/EC) affecting over 10,000 manufacturers and power generators in the European Union (EU), as an important step towards meeting the requirements set forth by the Kyoto Protocol. In the US, several regional initiatives that rely on cap-and-trade programs have been proposed to curb greenhouse gas emissions. Examples include the Regional Greenhouse Gas Initiative (RGGI) in the northeast, California's Global Warming Solutions Act (AB 32, 2006), and the Western Regional Climate Action Initiative (WRCAI) among states in the western US and the province of British Columbia.

Analogous to the manner in which industrial ecologists comprehensively assess life-cycle impacts of product systems, firms as well as policy-makers are now recognizing the need to holistically address environmental decisions and targets across various product life-cycle stages. Life Cycle Assessment (LCA) methods provide a systematic accounting of environmental burdens and impacts for a defined product system (Smith and Keoleian 2004). More often than not, the industrial ecologist's view of the product life-cycle is more extensive than that of an individual firm which might be just one of the many links in the value-adding chain from virgin material extraction to final product disposal or recovery. Recent environmental regulations in the EU, the US, Canada, Japan, Korea and elsewhere force individual companies to expand their view of the product life-cycle by making them responsible for environmental impacts beyond factory gates (for example, see Scheer and Rubik 2005). Indeed, Extended Producer Responsibility (EPR) programs such as the ELV directive seek to give producers appropriate incentives and signals concerning the life-cycle environmental impacts of their products. Notably, the EC is actively pursuing the broader concept of Integrated Product Policy (IPP). IPP seeks to minimize life-cycle

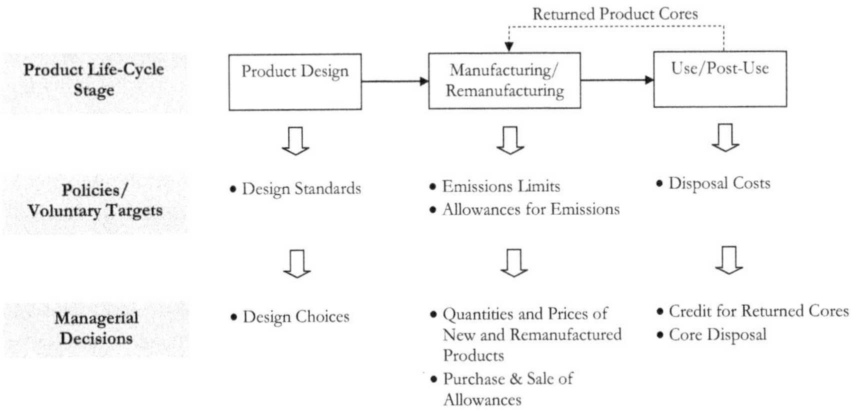

Fig. 14.1 Scope of the model

environmental impacts of products by combining instruments such as economic incentives, substance bans, voluntary agreements, environmental labeling, and product design guidelines, while at the same time emphasizing competitiveness concerns. IPP pilot projects involving various stakeholders including manufacturers, suppliers, governmental agencies, customers, and NGOs are currently underway (Environmental Resources Management 2005; Nokia Corporation 2005). Product-oriented policies such as IPP impact the value proposition of businesses and necessitate the incorporation of environmental aspects into strategic thinking. In its green paper on IPP, the EC recognizes the need to develop methodologies that enable a better understanding of the various drivers of environmental decisions by businesses (Tukker 2006; EC 2001).

Today it is clear that the connection between a firm's operational decisions and its environmental performance is immediate (Tukker and Jansen 2006). While environmental considerations often impose additional costs and constraints on production systems, they also open up new opportunities that, if properly exploited, can lead to better financial performance while also improving the firm's environmental impact. Joint operational and environmental decision-making requires the understanding and modeling of complex trade-offs, which in turn requires a rich and pliable framework capable of treating non-linear interactions (Bloemhof-Ruwaard et al. 1995). Senior managers in many industries lack such a framework and practical tools that can help them set priorities and make decisions that are both financially and environmentally sound. To address this need, we develop a non-linear mathematical programming model from a profit-maximizing firm's perspective, which can be tailored as a decision-support tool for firms facing environmental goals and constraints. Figure 14.1 outlines the scope of our model, including the specific product life-cycle stages that are captured.

Although we present a characteristic model for a specific industry context, the approach in itself is sufficiently flexible to accommodate a range of applications.

It allows for the treatment of a spectrum of regulatory instruments such as design standards, product take-back, disposal costs, emissions limits and charges, and environmental end-user fees, together with compliance options such as product design, investment in abatement technologies, procurement of emissions allowances, and adjustment of the product mix (including the mix of new and remanufactured products) or raw material inputs. It also allows for the consideration of key strategic aspects such as product pricing, consumer demand, competition and the cannibalization of new products by remanufactured products. From a methodological perspective, our work demonstrates how environmental targets and firms' compliance strategies can be modeled effectively using the classical operations research tool of mathematical programming. The model's ability to treat non-linear relationships makes it amenable to refinement and managerial application. The work is timely and pertinent to practitioners, many of whom are beginning to deal with complex tradeoffs in complying with environmental regulations at various stages in the production and consumption of their products.

14.2 The Need for a Holistic Model

While managerial decision-making is challenging in its own right, present day environmental policies force firms to address numerous additional tradeoffs. The subset of the product life-cycle relevant to managerial decision-making continues to expand as environmental policy-making increasingly attributes responsibilities for the life-cycle environmental impacts of products to manufacturers. We typify our approach using the specific context of diesel engine manufacturing and remanufacturing. Our model constructs are based on detailed interviews in the year 2008 with top managers from two leading competitors (labeled as Companies A and B in our discussions and model development) in the medium and heavy-duty diesel engine industry.

Our interviews followed a semi-structured format. We sent a set of questions to each interviewee in advance, and followed it up by a series of phone interviews during which we let the interviewees talk about our specific questions as well as any other related issues. Thereafter, we sent follow-up questions via email to have certain aspects clarified or expanded upon.

Although the companies interviewed have a sense as to the various types of interactions between operations planning and environmental factors, there is currently not a holistic effort in place to characterize the interactions and tie the factors together. Company A, for example, has a department devoted to assessing the environmental impacts of its products. However, there is limited meshing of the department's actions and operations planning decisions. As a result, the feeling is that monetary as well as environmental benefits are not fully realized. Environmental considerations are primarily tied into the design phase to ensure that legislated standards are met, but are not actively treated in operational planning decisions such as the product mix. For example, both companies have put voluntary corporate programs in place to reduce their carbon impact. Since remanufacturing requires significantly less energy than manufacturing new, it would make sense to tie remanufacturing decisions with voluntary commitments. A point worth noting is that Company B owns foundries in Europe that are affected by the ETS, which is aimed at limiting CO_2 emissions. However, environmental considerations do not currently drive remanufacturing decisions at Company B; remanufacturing is not pursued if the associated cost is expected to be more than a specified percentage of the cost of producing new.

With respect to the coordination of manufacturing and remanufacturing decisions, remanufacturing operations did not historically affect decisions related to new products at the companies interviewed. Although there is currently some level of coordination, the sense is that manufacturing and remanufacturing executives are heavily focused on their respective operations. For example, Company A recognizes the issue of new and remanufactured products competing for capacity. However, it lacks a practical tool to intelligently decide how scarce capacity should be allocated. Among the methods employed at Company A, is conducting low-volume assembly of new products at the remanufacturing facility. With respect to product pricing, the companies use price targets for remanufactured products (that is, the price of a remanufactured product is a specified percentage of the price of its new counterpart—currently in the range of 60–85%). However, the companies believe that pricing decisions should be made more rigorously. Additionally, both companies offer monetary incentives to drive return flows of used product cores. Company A, for example, believes that the current method of offering core credit is much less scientific than it needs to be. The sense is that the credit offered for cores should relate to inventory and planned production levels, but related decision-making tools are lacking.

14.3 Model Development

In this section, we progressively build the model by motivating and characterizing the various decisions, constraints, and inter-relationships involved in decision-making at the two companies interviewed. In developing our multi-product, multi-period manufacturing-remanufacturing model from a company's perspective, we use the subscript $i \in \{1 \ldots M\}$ to denote the index of the product and $t \in \{1 \ldots T\}$ to denote the time period (say, months or years). M indicates the number of products (say, within a product family) and T indicates the time horizon over which decisions are to be optimized. We use the subscripts n and r to refer to new and remanufactured products, respectively. The model development reflects numerous tradeoffs across product life-cycle stages—starting with product design, through production, to the end of the product's economic life. The functional forms used to model various relationships are structurally[1] consistent with the interviewed managers' experience. Specific values of the parameters of the functional specifications are currently unknown, but can be estimated with varying levels of effort (discussed further in Sect. 14.4).

[1] that is, convex/linear/concave, and increasing/decreasing.

14.3.1 *Product Design*

14.3.1.1 Product Performance

Diesel engine emissions and fuel economy standards are specified by legislation. The fuel mileage of the engine is a primary design attribute that directly impacts engine emissions. For exposition, we label this design attribute as *performance*. Let $Q_{i,\text{std}}$ denote the standard (or lower bound) of performance for product i prescribed by regulation, and let Q_{i0} denote the manufacturer's choice of performance of product i. The subscript 0 in Q_{i0} denotes that the design decision is made at the beginning of the planning horizon. Let Q_{i0}^C denote the performance of the competitor's corresponding product i. Consistent with the diminishing returns to design investments, we model the design cost of performance for product i as $[\xi_{1i} Q_{i0}]^2$, where $\xi_{1i} > 0$.

Decision 1 Determine the performance of product i: $Q_{i0} \geq 0$.

Constraint 1 Performance standard for product i: $Q_{i0} \geq Q_{i,std}$.

Cost 1 Design cost of performance for product i: $[\xi_{1i} Q_{i0}]^2$.

14.3.1.2 Product Remanufacturability

Although existing product take-back laws do not currently affect either company's products, there are concerns that sooner or later entire product families will be subject to take-back laws. Despite the absence of take-back requirements, both companies have been collecting engines from the marketplace and remanufacturing them due to the associated value in doing so. Engine remanufacturing requires between 80–90% less energy than manufacturing new, and the revenue generated in the aftermarket by remanufacturing is attractive. Remanufacturing also costs less due to the reuse of materials and the relatively lower processing costs. Both Company A and Company B have explicit Design for Remanufacturability (DfR) criteria, including designing wall thicknesses so that the engine block can be milled or bored down during the remanufacturing process, allowing for the building up of material where necessary, and in general designing products that are easy to clean and machine during remanufacturing. Other DfR efforts include informational tools such as ECUs (Engine Control Units) and fatigue clocks built into the engine, which record duty cycles and other pertinent information to facilitate remanufacturing.

Therefore, in addition to the design choice of performance described earlier, we also explicitly model the choice of *remanufacturability*. Let $\Theta_{i0} \in [0, 1 - \Theta_{iB}]$ denote the manufacturer's choice of remanufacturability of product i, representing the fraction (say, by weight) of a core of product i that can be effectively remanufactured beyond the inherent level (that is, absent any investment in remanufacturability) Θ_{iB}. For example, when greater remanufacturability is built into the design of the engine, major castings or forgings such as blocks, cylinder heads, crankshafts, and connecting rods can be more effectively remanufactured. We model the design cost of remanufacturability for product i as $[\xi_{2i} \Theta_{i0}]^2$, where $\zeta_{2i} > 0$.

Decision 2 Determine the remanufacturability of product i: $\Theta_{i0} \geq 0, \leq 1$.

Cost 2 Design cost of remanufacturability for product i: $[\xi_{2i}\Theta_{i0}]^2$.

14.3.2 Manufacturing/Remanufacturing Costs

Remanufacturability often adversely impacts unit production costs for new engines. For example, with the use of metal removal technologies to remanufacture engine components, the extra casting or forging thicknesses result in increased material content and, hence, increased costs. Additionally, to capture the increasing costs in relation to performance, we model the unit cost of manufacturing a new product i in period t: as $c_{itn} = \tilde{c}_{itn0} + \tilde{c}_{itn} Q_{i0}^2 \Theta_{i0}$, where $\tilde{c}_{itn0}, \tilde{c}_{itn} > 0$. We denote f_{itn} as the fixed cost of manufacturing new product i in period t.

On the other hand, for engines within the same displacement family, remanufacturing costs are not adversely affected by the engine's performance. Since the variable cost of remanufacturing decreases with remanufacturability, we model the unit cost of remanufacturing product i in period t as $c_{itr} = \tilde{c}_{itr0} + \tilde{c}_{itr}[1 - (\Theta_{i0} + \Theta_{iB})^2]$, where $\tilde{c}_{itr0}, \tilde{c}_{itr} > 0$. We denote f_{itr} as the fixed cost of remanufacturing product i in period t. Due to its labor intensity, remanufacturing has lower fixed costs than manufacturing new.

Relationship 1 Unit cost of manufacturing a new product i in period t: $c_{itn} = \tilde{c}_{itn0} + \tilde{c}_{itn} Q_{i0}^2 \Theta_{i0}$.

Relationship 2 Unit cost of remanufacturing product i in period t: $c_{itr} = \tilde{c}_{itr0} + \tilde{c}_{itr}[1 - (\Theta_{i0} + \Theta_{iB})^2]$.

14.3.3 Product Mix

The product mix decision involves determining the quantities of new and remanufactured products to produce in each period. The choice of production quantities is intricate since products have different manufacturing and remanufacturing costs, compete for capacity, are subject to different demand parameters, are associated with different levels of emissions, face different core disposal costs (or salvage values), have different economic lives, and have different inventory holding and backordering costs. As mentioned in Sect. 14.3.1.2, remanufacturing is attractive due to its profit potential. However, remanufacturing also has the fundamental environmental benefit of lower energy use. About 95% of CO_2 emissions during engine manufacturing can be attributed to energy consumption, and remanufacturing requires between 80–90% less energy than manufacturing new. Hence, remanufacturing should be favored towards meeting either voluntary or mandated limits on CO_2 emissions.

We denote X_{it} and Y_{it}, respectively, as the number of units of product i manufactured and remanufactured in period t. Let h_{itn}, h_{itr}, and h_{itc} respectively denote the unit inventory holding costs of new product i, remanufactured product i, and cores of product i in period t. Let u_{itn} and u_{itr} respectively denote the unit cost of backordering customer demand for new and remanufactured product i in period t. Additionally, current engine design tools are reasonably accurate in predicting the economic life of an engine; we denote τ_i as the economic life of product i (measured in time periods).

Non-remanufacturable components of cores, as well as entire cores that are decided not to be remanufactured at all (perhaps due to holding costs), undergo both scrapping as well as disposal of certain parts. Often, the cast iron, steel, and aluminium in a core can be profitably scrapped. Disposal costs are incurred for dirt that comes off the core and for certain consumable parts such as gaskets. Let ρ_{it} denote the net unit cost of disposing of (or, equivalently, the net salvage value per unit of) cores of product i in period t. Let W_{it} denote the number of entire cores of product i disposed of in period t.

Decision 3 Determine the production quantities of new and remanufactured products in period t: $X_{it}, Y_{it} \geq 0$.

Decision 4 Determine the number of cores to be disposed of in period t: $W_{it} \geq 0$.

Cost 3 Total cost of manufacturing new product i in period t: $\delta_{itn} f_{itn} + c_{itn} X_{it}$, where $\delta_{itn} = 1$ if $X_{it} > 0$; 0 otherwise.

Cost 4 Total cost of remanufacturing product i in period t: $\delta_{itr} f_{itr} + c_{itr} Y_{it}$, where $\delta_{itr} = 1$ if $Y_{it} > 0$; 0 otherwise.

14.3.4 Manufacturing/Remanufacturing Capacity

The two principal types of capacity in manufacturing and remanufacturing are machining and assembly. Although the assembly operation in remanufacturing is more labor intensive than in manufacturing new, some assembly capacity can be shared across remanufacturing and manufacturing. However, this is not typical of machining capacity. We denote K_{at} as the total assembly capacity available in period t, K_{mtn} as the machining capacity available for manufacturing new in period t, and K_{mtr} as the machining capacity available for remanufacturing in period t.

New and remanufactured versions of a product consume different amounts of capacity. For example, remanufacturing requires a smaller amount of machining capacity as compared to manufacturing new. Therefore, we denote k_{ain} and k_{air}, respectively, as the manufacturing and remanufacturing assembly capacities consumed per unit of product i. Similarly, k_{min} and k_{mir} denote the machining capacities consumed per unit of product i.

Constraint 2 Assembly capacity available in period t: $\sum_{i=1}^{M} [k_{ain} X_{it} + k_{air} Y_{it}] \leq K_{at}$.

Constraint 3 Machining capacity available for manufacturing a new product in period t: $\sum_{i=1}^{M} k_{min} X_{it} \leq K_{mtn}$.

Constraint 4 Machining capacity available for remanufacturing in period t: $\sum_{i=1}^{M} k_{mir} Y_{it} \leq K_{mtr}$.

14.3.5 Pricing and Demand

Price adjustments are made on a periodic basis at the companies interviewed (annually, in the case of Company A). Exceptions to this periodicity include adjustments to compensate for extraordinary fluctuations in raw material costs. We denote P_{itn} and P_{itr}, respectively, as the prices of new and remanufactured product i in period t.

Price differentials between new and remanufactured engines are known to affect respective customer demands. Also, different markets are known to have varying levels of access to and acceptability of remanufactured engines. Performance is a major driver of demand in developed markets as opposed to price in developing markets.[2] Interestingly though, customers in developed countries are known to embrace remanufactured engines more than customers in developing countries.[3] We model the demands for new and remanufactured versions of a product in each period as functions of prices of the versions, product performance, and the price $\left(P_{it}^{C}\right)$ and performance $\left(Q_{i0}^{C}\right)$ of the competing product offered by the competition.

Decision 5 Determine the prices of new and remanufactured product i in period t: $P_{itn}, P_{itr} \geq 0$.

Relationship 3 Demand for new product i in period t: $d_{itn} = a_{itn} - b_{itn}P_{itn} + b_{itnr}P_{itr} + b_{itn}^{C}P_{it}^{C} + b_{itnq}Q_{i0} - b_{itnq}^{C}Q_{i0}^{C} \geq 0$; Demand for remanufactured product i in period t: $d_{itr} = a_{itr} - b_{itr}P_{itr} + b_{itrn}P_{itn} + b_{itr}^{C}P_{it}^{C} + b_{itrq}Q_{i0} - b_{itrq}^{C}Q_{i0}^{C} \geq 0$. All $a_{[]}$ (market potentials) and $b_{[]}$ (sensitivity parameters) ≥ 0.

Relationship 3 captures the cannibalization effect that remanufactured products may have on the demand for new products. Parameters b_{itnr} represent sensitivity of demand for new products to the prices of remanufactured products in period t. Similarly, parameters b_{itrn} represent sensitivity of demand for remanufactured products to the prices of new products in period t. Thus, new and remanufactured products are partial substitutes. In general, demand for a product is more sensitive to changes in its own price than to changes in prices of substitutes (that is, $b_{itn} > b_{itnr}$ and $b_{itr} > b_{itrn}$, $\forall i, t$). As mentioned above, markets differ in their preferences and price sensitivities for new and remanufactured products.

Revenue 1 Revenues from new and remanufactured product i in period t: $P_{itn}d_{itn} + P_{itr}d_{itr}$.

Relationship 4 Inventory position[4] of new product i at end of period t: $IP_{itn} = IP_{i,t-1,n} + X_{it} - d_{itn}$; Inventory of new product i at end of period t: $I_{itn} = [IP_{itn}]^{+}$; Backordered quantity of new product i at end of period t: $\sigma_{itn} = [IP_{itn}]^{-}$.

---

[2] However, there is a sense that customers in developing markets are now gravitating towards similar preferences as customers in developed markets.

[3] A reason mentioned by one of the managers interviewed is that customers in developed markets value uptime and the comparable performance afforded by remanufactured engines, while in developing markets where uptime is less critical and labor costs are low, repair is preferred as a less expensive alternative to buying a remanufactured engine.

[4] A positive value of IP_{itn} or IP_{itr} indicates inventory on hand whereas a negative value indicates a backordered quantity.

Relationship 5 Inventory position of remanufactured product i at end of period t: $IP_{itr} = IP_{i,t-1,r} + Y_{it} - d_{itr}$; Inventory of remanufactured product i at end of period t: $I_{itr} = [IP_{itr}]^{+}$; Backordered quantity of remanufactured product i at end of period t: $\sigma_{itr} = [IP_{itr}]^{-}$.

Cost 5 Costs of backordering new and remanufactured product i in period t: $u_{itn}\sigma_{itn} + u_{itr}\sigma_{itr}$.

Relationship 6 Sales of new product i in period t: $S_{itn} = I_{i,t-1,n} + X_{it} - I_{i,t,n}$; Sales of remanufactured product i in period t: $S_{itr} = I_{i,t-1,r} + Y_{it} - I_{i,t,r}$.

14.3.6 Core Credit

Return flows of end-of-life engine cores can be influenced by monetary incentives. We denote Ψ_{it} as the credit offered per returned core of product i in period t.[5] By offering such incentives, the companies are better able to secure control over cores, which are regarded as the "life-blood" of remanufacturing operations and which competing third-party engine remanufacturers and rebuilders simultaneously seek out. Thus, end-of-life core returns are sensitive to the core credit offered. We model core returns of product i in period t as an increasing function of the core credit offered.

Return flows of cores involve fixed costs related to (i) facilities for transporting, receiving, inspecting, and storing cores; (ii) the materials handling infrastructure; and, (iii) the information systems infrastructure required for inventory management as well as for the transactional aspects of receiving, inspecting, and issuing credit for returned cores. We denote f_{itc} and c_{itc}, respectively, as the fixed and variable costs of collecting cores of product i in period t.

Decision 6 Determine the credit per unit of returned core of product i in period t: $\Psi_{it} \geq 0$.

Relationship 7 Returns of cores of product i in period t: $R_{it} = \lambda_{it}\Psi_{it}[S_{i,t-\tau_i,n} + S_{i,t-\tau_i,r}]$, where $\lambda_{it} \in [0,1]$ reflects the sensitivity of core returns to the credit offered.[6]

---

[5] An alternative way by which the companies interviewed influence the returns of cores is through an upfront "core charge" at the time of sale, which is refunded when the core is returned after use. This is analogous to a deposit-refund system, which has been successfully employed by various countries to encourage the recycling of products such as beverage containers and automotive batteries.

[6] For expositional convenience, we assume that in any given period, only cores sold τ_i periods ago will return for remanufacture, where τ_i is the economic life of product i. A more general formulation could allow for core returns from sales in any prior period. We also noted from the interviews that there is limited difference in the economic lives of new and remanufactured versions of an engine, although there is an upper bound on the number of times a core can be remanufactured. For over the road truck engines, it is common to remanufacture a core up to five times.

Constraint 5 Possible returns of cores of product i in period t: $R_{it} \leq S_{i,t-\tau_i,n} + S_{i,t-\tau_i,r}$.

Cost 6 Total cost incurred for returned cores of product i in period t: $\delta_{itc} f_{itc} + (\Psi_{it} + c_{itc}) R_{it}$; $\delta_{itc} = 1 \quad \text{if} \quad R_{it} > 0$; 0 otherwise.

Cost 7 Total disposal cost incurred for cores of product i in period t: $\rho_{it}[\{1-(\Theta_{i0}+\Theta_{iB})\}Y_{it} + W_{it}]$ (also see Decision 4)

Relationship 8 Inventory of cores of product i at end of period t, after core returns, disposal, and remanufacture: $I_{itc} = I_{i,t-1,c} + R_{it} - W_{it} - Y_{it}$.

Constraint 6 Number of cores of product i that can possibly be disposed of in period t: $W_{it} \leq I_{i,t-1,c} + R_{it}$.

Constraint 7 Number of cores of product i that can possibly be remanufactured in period t: $Y_{it} \leq I_{i,t-1,c} + R_{it} - W_{it}$.

Cost 8 Cost of holding inventories of new product i, remanufactured product i, and cores of product i in period t: $h_{itn} I_{itn} + h_{itr} I_{itr} + h_{itc} I_{itc}$.

14.3.7 Limits on Emissions

The principal pollutant attributable to engine manufacturing and remanufacturing is CO_2 from the energy consumed. Both companies interviewed have set voluntary targets of reductions in carbon emissions attributable to their facilities. Specifically, Company A has committed to reduce CO_2 emissions due to its facilities by 25% (normalized by sales) by the year 2010 as compared to 2005. Company B has recently started measuring carbon impact and has self-imposed goals on reducing this impact. Moreover, as noted earlier, Company B owns foundries in Europe that are impacted by the EU-wide cap-and-trade program (ETS), in which affected companies are required to account for CO_2 emissions with the requisite number of allowances. Let e_{in} and e_{ir} respectively denote the CO_2 emissions attributable to the manufacture and remanufacture of a unit of product i. Let l_t denote the voluntary limit on such emissions in period t.[7] Let B_t and V_t respectively denote the number of allowances purchased and sold in period t. Let η_t (assumed to be $> l_t$) denote the number of allowances available for purchase in the market, and ϕ_t denote the market price of an allowance in period t.

[7] Company B is not constrained by the market availability of allowances and also does not face a mandated limit on CO_2 emissions. However, under the ETS, all CO_2 emissions have to be accounted for by allowances that have market value. From our model, the shadow price associated with the voluntary emissions constraint can be used to assess the impact on profitability of voluntarily limiting CO_2 emissions. The model can be easily adapted to accommodate a mandated limit and/or constrained market availability of allowances, if applicable.

Relationship 9 Total CO_2 emissions attributable to manufacturing and remanufacturing in period t, $E_t = \sum_{i=1}^{M} (e_{in} X_{it} + e_{ir} Y_{it})$.

Constraint 8 Emissions limit: $E_t \le l_t$.

Decision 7 Determine the number of allowances to be purchased in period t: $B_t \ge 0$, and the number of allowances sold in period t: $V_t \ge 0$.

Cost 9 Cost of allowances purchased in period $t = \phi_t B_t$.

Constraint 9 Market availability of allowances in period t: $B_t \le \eta_t$.

Relationship 10 Inventory of allowances at end of period t: $\beta_t = \beta_{t-1} + B_t - E_t - V_t$.

Constraint 10 Allowances on hand in period t: $E_t \le \beta_{t-1} + B_t$.

Constraint 11 Number of allowances that can possibly be sold in period t: $V_t \le \beta_{t-1} + B_t - E_t$.

Revenue 2 Revenue from sale of allowances in period t: $\phi_t V_t$.

14.3.8 Initial and Terminal Conditions

For model completeness, we specify the following null initial and terminal conditions. In operational use, the right hand sides of the initial conditions would take on the actual values at the time of model implementation while the terminal conditions would reflect acceptable bounds for the end of the planning horizon.

Initial Conditions (Zero Inventories and Zero Core Returns, Disposal, and Remanufacturing):

$$IP_{i,0,n}, IP_{i,0,r} = 0; \quad i = 1, 2, \dots, M$$
$$I_{ikc}, R_{ik}, W_{ik}, Y_{ik} = 0; \quad i = 1, 2, \dots, M; \; k = 1, 2, \dots, \tau_i$$
$$\beta_0 = 0$$

Terminal Conditions (Zero Backorders):

$$\sigma_{iTn}, \sigma_{iTr} = 0; \quad i = 1, 2, \dots, M$$

14.3.9 Complete Model

In this section, we present the complete model that assembles together the revenues, costs, relationships, constraints, and conditions identified in Sects. 14.3.1

Model Component	Operational Elements	Environmental Elements	Strategic Elements
Decision Variables	• Quantities of new and remanufactured products	• Design choices (*Performance & Remanufacturability*) • Disposal of cores • Number of allowances to purchase or sell	• Prices of new and remanufactured products • Credit for core returns
Parameters requiring relatively lesser additional effort for estimation	• Fixed and variable costs of manufacturing, remanufacturing, and core collection • Economic life of products • Capacities consumed per unit of new and remanufactured products • Inventory holding costs for new and remanufactured products, and product cores	• Disposal costs • Market prices of allowances	
Parameters requiring relatively greater additional effort for estimation	• Backordering costs for new and remanufactured products	• Design costs for performance and remanufacturability • Emissions attributable to the manufacture and remanufacture of a unit of a product	• Demand parameters: Market potential, Price sensitivity, Substitutability of new and remanufactured counterparts, Sensitivity to product performance, Sensitivities to price and performance of competing products • Sensitivity of core returns to core credit
Constraints	• Available manufacturing and remanufacturing capacities • Inventory balance for new and remanufactured products, and product cores	• Design standard of performance • Voluntarily specified emissions limits • Allowances available for purchase • Allowance inventory balance	

Fig. 14.2 Model elements

through 14.3.8. Subsequent to the initial design choices of performance and remanufacturability, we assume the following sequence of activities in each decision period. Collected cores become available to the manufacturer, cores from inventory may be disposed of, new and remanufactured products are produced, non-remanufacturable components are disposed of, emissions allowances are purchased, applied and sold, prices for new and remanufactured products are offered and demands are realized, and finally, inventories and associated costs are assessed. The elements captured in the model are summarized in Fig. 14.2.

In words, the manufacturer maximizes its discounted total profit (we denote the discount factor as $\alpha \in (0, 1)$) across all products and time periods, expressed as: Revenue from sales of new and remanufactured products + Net Revenue from allowances—Cost of product design—Cost of manufacturing and remanufacturing—Cost of core collection—Cost of holding inventory—Cost of backordering customer demand—Cost of disposal of cores and non-remanufacturable components. The key factors that constrain profit maximization can be classified under two categories—operational and environmental. Operational factors include manufacturing and remanufacturing capacities (Constraints 2, 3, and 4), while environmental factors include design standards (Constraint 1), availability of emissions allowances (Constraint 9), voluntarily specified emissions limits (Constraint 8), and emissions caps (Constraint 10).

Maximize:

$$\begin{aligned}&\sum\nolimits_{t=1}^{T}\sum\nolimits_{i=1}^{M}\alpha^{t}[P_{itn}d_{itn}+P_{itr}d_{itr}]+\sum\nolimits_{t=1}^{T}\alpha^{t}\phi_{t}[V_{t}-B_{t}]\\&-\sum\nolimits_{i=1}^{M}[(\xi_{1i}Q_{i0})^{2}+(\xi_{2i}\Theta_{i0})^{2}]\\&-\sum\nolimits_{t=1}^{T}\sum\nolimits_{i=1}^{M}\alpha^{t}[\delta_{itn}f_{itn}+c_{itn}X_{it}+\delta_{itr}f_{itr}\\&+c_{itr}Y_{it}+\delta_{itc}f_{itc}+(\Psi_{it}+c_{itc})R_{it}\\&+h_{itn}I_{itn}+h_{itr}I_{itr}+h_{itc}I_{itc}\\&+u_{itn}\sigma_{itn}+u_{itr}\sigma_{itr}+\rho_{it}[\{1-(\Theta_{i0}+\Theta_{iB})\}Y_{it}+W_{it}]]\end{aligned}$$

Subject to Constraints:

$$Q_{i0}\geq Q_{i,(\text{std})};\qquad i=1,2,\ldots,M\quad\text{(Constraint 1)}$$

$$\sum_{i=1}^{M}[k_{ain}X_{it}+k_{air}Y_{it}]\leq K_{at};\qquad t=1,2,\ldots,T\quad\text{(Constraint 2)}$$

$$\sum_{i=1}^{M}k_{min}X_{it}\leq K_{mtn};\qquad t=1,2,\ldots,T\quad\text{(Constraint 3)}$$

$$\sum_{i=1}^{M}k_{mir}Y_{it}\leq K_{mtr};\qquad t=1,2,\ldots,T\quad\text{(Constraint 4)}$$

$$R_{ik}\leq S_{i,k-\tau_i,n}+S_{i,k-\tau_i,r};\quad i=1,2,\ldots,M;\ k=\tau_i+1,\tau_i+2,\ldots,T\quad\text{(Constraint 5)}$$

$$W_{ik}\leq I_{i,k-1,c}+R_{ik};\quad i=1,2,\ldots,M;\ k=\tau_i+1,\tau_i+2,\ldots,T\quad\text{(Constraint 6)}$$

$$Y_{ik}\leq I_{i,k-1,c}+R_{ik}-W_{ik};\quad i=1,2,\ldots,M;\ k=\tau_i+1,\tau_i+2,\ldots,T\quad\text{(Constraint 7)}$$

$$E_t\leq l_t;\qquad t=1,2,\ldots,T\quad\text{(Constraint 8)}$$

$$0\leq B_t\leq\eta_t;\qquad t=1,2,\ldots,T\quad\text{(Constraint 9)}$$

$$E_t\leq\beta_{t-1}+B_t;\qquad t=1,2,\ldots,T\quad\text{(Constraint 10)}$$

$$0\leq V_t\leq\beta_{t-1}+B_t-E_t;\qquad t=1,2,\ldots,T\quad\text{(Constraint 11)}$$

$$
\begin{aligned}
0 \leq \Theta_{i0} \leq 1 - \Theta_{iB}; \quad & i = 1,2,\ldots,M \\
\Psi_{it}, X_{it}, P_{itn}, P_{itr}, d_{itn}, d_{itr} \geq 0; \quad & i = 1,2,\ldots,M;\ t = 1,2,\ldots,T \\
W_{ik}, Y_{ik} \geq 0; \quad & i = 1,2,\ldots,M;\ k = \tau_i + 1, \tau_i + 2, \ldots, T \\
I_{ikc}, R_{ik}, W_{ik}, Y_{ik} = 0; \quad & i = 1,2,\ldots,M;\ k = 1,2,\ldots,\tau_i \\
IP_{i,0,n}, IP_{i,0,r}, \sigma_{iTn}, \sigma_{iTr} = 0; \quad & i = 1,2,\ldots,M \\
\beta_0 = 0 \quad &
\end{aligned}
$$

14.4 Implementation

In this section, we discuss issues with implementing the modeling approach presented in Sect. 14.3. While it is clear from the discussion in Sect. 14.2 that the two companies interviewed feel the need for holistic decision models that mesh various interactions together, neither company currently undertakes such an effort due to a lack of appropriate tools and data. To the best of our knowledge, the same is true in other industries.

Our primary goal in this chapter is to demonstrate how a decision model based on a variety of observed interactions can be developed in a systematic manner. In practice, our model would be modified to capture the specific strategic, tactical, and environmental considerations of a given firm and populated with firm-specific data. The approach requires the characterization and estimation of several relationships and parameters, some of which are known and others which require data that most firms do not normally collect—at least systematically. Based on the information gathered from the interviews, we classify the model's parameters in terms of relative additional effort required in estimating them (see Fig. 14.2). Certain data that would help in estimating specific parameters are likely to currently exist, although in a fragmented form across organizational functions. Fixed and variable costs of manufacturing and remanufacturing, and inventory holding costs, can be obtained from standard accounting systems, but would require some effort if such data is not already used routinely. The design function would have reasonable estimates as to the expected economic lives of products. Manufacturing and remanufacturing capacities are reasonably known to the manufacturing/remanufacturing functions as are disposal costs (or salvage values) of cores. The finance and accounting functions would have data on emissions allowances, where applicable.

Estimating the remaining relationships and associated parameters would require creativity and effort, since the necessary data is unavailable even to people charged with making related decisions. The feeling at the companies interviewed is that there are often situations where assumptions that feed decision-making are made without supporting data. For example, in Company A, some steps with respect to understanding demand parameters are being undertaken, but the belief is that much more improvement is possible with respect to assessing how customers value various attributes such as price, performance, and the offerings by the competition. In addition, structured efforts will be required to assess the extent of cannibalization of new products by remanufactured products, the opportunity costs of not being able to meet customer demand, and the sensitivity of core returns to the credit offered. The marketing functions at the companies interviewed currently have a sense as to the general nature of these relationships, but formal assessments are lacking.

The model developed in this chapter highlights the importance of LCAs in their currently untapped ability to inform managerial decision-making. Of specific relevance to the model developed in this chapter is the LCA model for remanufactured engines by Smith and Keoleian (2004), who provide numeric estimates of air emissions, solid waste generation, material consumption, and real cost savings attributable to remanufacturing. Such information in LCAs can be valuable in estimating the model's parameters, thus enabling the bridging of operational and environmental decisions, for example, the mix between new and remanufactured products so as to meet environmental targets.

14.5 Model Solution: Illustration

As discussed in Sect. 14.4, it is currently impossible to use real data to populate the model. However, to demonstrate the importance of considering environmental factors in making managerial decisions, we populate and solve the model with illustrative data for two products over a time horizon of ten periods. Appendix 1 provides an overview of the characteristics of products 1 and 2 assumed in the base data. For clarity, we present results in the form of graphs in Appendix 2. The graphs are annotated with corresponding variations of the base data.

A variety of commercially-available optimization software exists for solving non-linear optimization problems, allowing for settings that typically trade-off speed with solution quality (see NEOS 2011 and Fourer 2011 for detailed discussions). We solve the model using the demo versions of AMPL and the MINOS solver.[8]

We briefly reflect, in turn, on how the consideration of three environmental factors—the emissions limit, the market price of allowances, and the core disposal cost, impacts decisions pertaining to the product mix (Fig. 14.3), product design (Figs. 14.4 and 14.5), and the credit offered to induce returns of cores (Fig. 14.6). In each of the Figures, the bar(s) on the extreme right, labeled as "All Factors Considered", represent the optimal decisions when all the three environmental factors are considered whereas the bars on the left represent the optimal decisions when one of the three factors, in turn, is not considered in the decision-making.

Emissions Limit To focus on the importance of factoring in the emissions limit in the decision-making, we contrast the situation where the companies do not consider the emissions limit (labeled as "Emissions Limit Not Considered" in Figs. 14.3–14.6) to the situation where all factors are considered. With the emissions limit

[8] AMPL® stands for "A Modeling Language for Mathematical Programming" (Fourer 2003). Demo versions of AMPL and MINOS are available at http://www.ampl.com

becoming an active constraint, the product mix changes significantly, with the overall production level decreasing and the level of remanufacturing activity (relative to manufacturing new) increasing.[9] The increase in remanufacturing is due to the lower emissions as compared to manufacturing new. As per our data, when emissions are not considered in operations planning, both manufacturing and remanufacturing are almost entirely focused on the more profitable product 2 and, therefore, significant investments are made in the performance and remanufacturability of product 2. However, with the emissions constraint becoming active, product 1, which has relatively lower emissions during production as compared to product 2, also becomes attractive to produce, with the remanufacturability for product 1 increasing relative to the unconstrained emissions case. Since we model the demand for a product as increasing in its performance, the prices of both new and remanufactured versions of product 1 are respectively greater when the emissions limit is not considered. Lastly, with emissions becoming constrained, the credit offered for cores increases since remanufacturing is more attractive under an active emissions constraint.

Market Price of Allowances To focus on the importance of factoring in the price of emissions allowances, we contrast the situation where the companies do not consider the price of allowances (labeled as "Allowance Price Not Considered" in Figs. 14.3–14.6) to the situation where all factors are considered. In other words, we contrast the optimal solution where allowances are regarded as costless from an operations planning standpoint (current industry practice) to the situation where allowances have a non-zero cost.[10] Similar to the discussion for the emissions limit above, when the price of allowances is considered, the product mix changes, with the level of remanufacturing activity for product 1 (relative to new manufacture) increasing. This increase is due to the lower emissions attributable to remanufacturing product 1 as compared to manufacturing new and correspondingly lower total cost incurred for emissions allowances. The reason for the slight decrease in remanufacturing activity for product 2 is the following. When the price of allowances is ignored, the remanufacturability of the more profitable product 2 is greater than that of product 1. However, with allowances becoming costly, product 1, which has relatively lower emissions during production as compared to product 2, becomes attractive for remanufacturing, with the remanufacturability for product 1 increasing and the remanufacturability of product 2 decreasing, relative to the zero allowance price case. Since in our model the variable cost of remanufacturing increases with a lack of remanufacturability, the level of remanufacturing for product 2 decreases when the market price of allowances is factored in the decision-making. Lastly, with allowances becoming expensive, the credit offered for cores of both products increases (marginally though for product 2) so as to facilitate remanufacturing activity.

Disposal Costs Finally, to focus on the importance of factoring in the cost of disposing of non-remanufacturable portions of returning cores, we contrast the situation

[9] A core may return several times for remanufacture.

[10] The emissions constraint is active in both cases.

where the companies do not consider the disposal cost of cores (labeled as "Disposal Cost Not Considered" in Figs. 14.3–14.6) to the situation where all factors are considered. In other words, we contrast the optimal solution where disposal is costless from an operations planning standpoint (current industry practice) to the situation where disposal is costly.[11] Similar to the discussion for the market price of allowances, when the cost of disposal is considered, the level of remanufacturing (relative to new manufacture) increases for product 1 but decreases for product 2. Interestingly, when core disposal is regarded as costless, the incentive to invest in product remanufacturability diminishes considerably—to the extreme in our illustrative data—since any non-remanufacturable portions of cores can be disposed of without cost.[12] Also, with disposal becoming costly, the increase in product remanufacturability justifies an increased credit for cores so as to facilitate remanufacturing activity.[13]

As the above results illustrate, joint operational and environmental decision-making involves complex and important trade-offs that are not currently being captured in practice. Without a decision-making tool capable of capturing these trade-offs holistically, firms are at risk of making suboptimal decisions with adverse consequences for both firm profitability as well as environmental outcomes.

14.6 Conclusion

The need for explicit consideration of environmental costs and constraints within operations planning is becoming critical to corporate management. An increasing number of firms across countries face environmental regulations at various stages in the production and consumption of their products. The non-linear programming approach posited in this chapter can treat key managerial tradeoffs starting with product design, through production, to the end of the product's economic life, and across multiple time periods and products. Although we present a characteristic model based on detailed interviews with top managers from two leading competitors in the medium and heavy-duty diesel engine industry (with the possibility that certain model elements may be context-specific, such as shared production capacity between new and remanufactured engines, voluntary emissions limits, permits for emissions, and various functional relationships), the approach in itself is flexible enough to accommodate a range of applications. In particular, the ability to treat non-linear relationships makes our approach capable of refinement and managerial application. It allows us to treat a spectrum of environmental considerations such as design

[11] Again, the emissions constraint is again active in cases.

[12] When the design choice of remanufacturability equals zero for product i, the level of remanufacturability is just the inherent level Θ_{iB}, implying that a remanufactured product i is almost entirely rebuilt.

[13] The companies interviewed do not currently face mandated product take-back. Therefore disposal costs are only incurred for the non-remanufacturable portions of only those cores that return for remanufacture.

standards and emissions targets, as well as strategic aspects such as product pricing, competition and the cannibalization of new products by remanufactured products. The work is timely and pertinent to practitioners, many of whom are just beginning to deal with complex tradeoffs in meeting environmental targets.

We recognize that the effort involved in characterizing the nature of relationships and associated parameters within the model is significant, but possible through the implementation of creative organizational and informational structures to systematically collect relevant data. We hope that this chapter successfully sheds light on the need to not only involve all departments within a business towards making sensible economic and environmental decisions, but also for industrial ecologists and business managers to work together in order to implement meaningful decision models that can have definite economic and environmental impact.

Acknowledgements This chapter is a summarized version of: Subramanian, R, B. Talbot, S. Gupta. 2010. An Approach to Integrating Environmental Considerations within Managerial Decision Making. *Journal of Industrial Ecology* 14(3) 378–398.

14.7 Appendix 1: Product Features Assumed in the Base Data

Following is a brief overview of the characteristics of products 1 and 2 assumed in the numerical illustration:

- Product 2 is more "complex" and requires more manufacturing and remanufacturing capacity per unit than product 1.
- Fixed and variable costs of manufacturing and remanufacturing are respectively higher for product 2 than for product 1. Also, holding and backordering costs are higher for product 2.
- Overall, customers of product 2 are less price sensitive and more quality conscious than customers of product 1. Customers of product 1 are less sensitive to the credit offered to induce core returns.
- The market sizes for both new and remanufactured product 1 are respectively larger than those for product 2.
- Emissions attributable to manufacturing and remanufacturing are respectively higher for product 2 than for product 1.
- The performance standard is higher for product 2.

14.8 Appendix 2: Illustrative Results

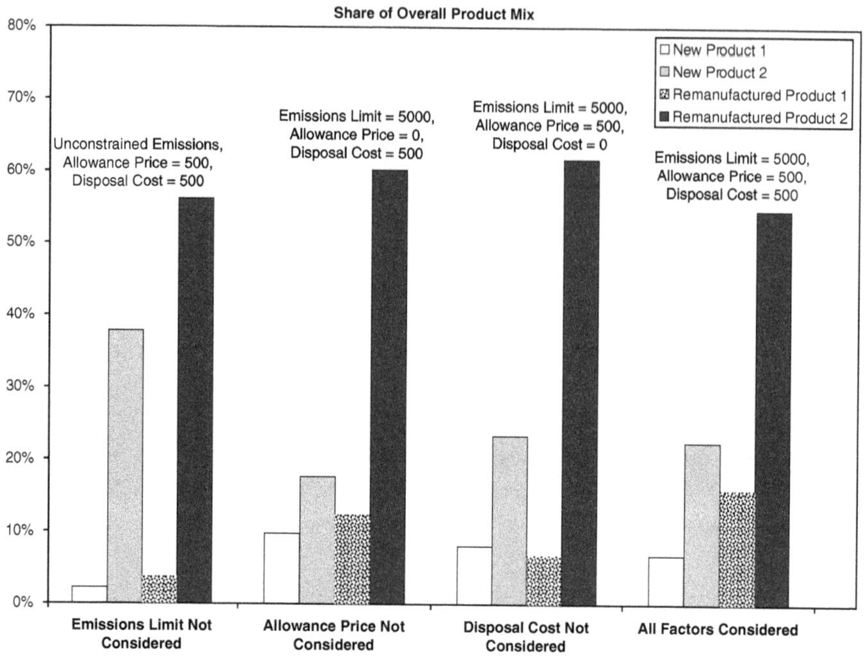

Fig. 14.3 Effect of environmental factors on product mix decision

$$(\text{Plotted Values} = \frac{\sum_{t=1}^{10} X_{1t}}{\sum_{i=1}^{2}\sum_{t=1}^{10}(X_{it}+Y_{it})}, \frac{\sum_{t=1}^{10} X_{2t}}{\sum_{i=1}^{2}\sum_{t=1}^{10}(X_{it}+Y_{it})}, \frac{\sum_{t=1}^{10} Y_{1t}}{\sum_{i=1}^{2}\sum_{t=1}^{10}(X_{it}+Y_{it})}, \frac{\sum_{t=1}^{10} Y_{2t}}{\sum_{i=1}^{2}\sum_{t=1}^{10}(X_{it}+Y_{it})})$$

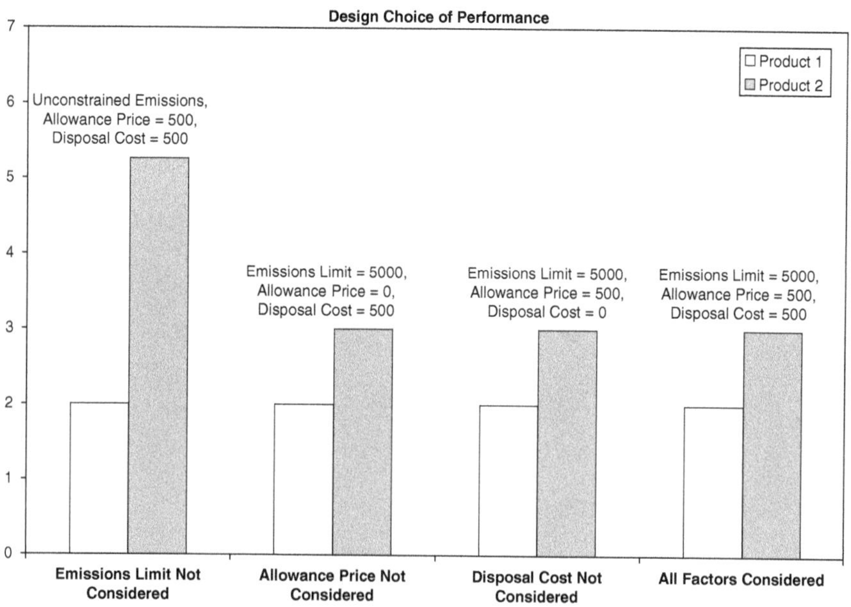

Fig. 14.4 Effect of environmental factors on design choice of performance (Plotted Values = Q_{10}, Q_{20})

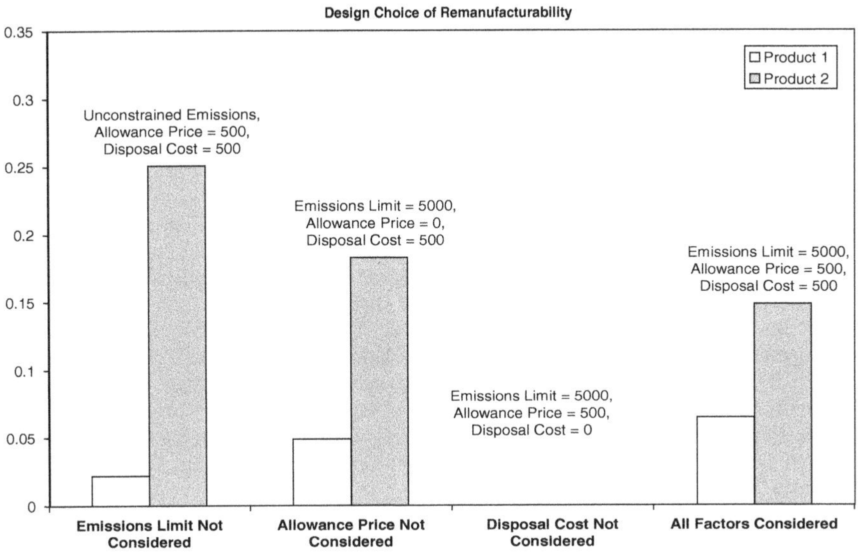

Fig. 14.5 Effect of environmental factors on design choice of remanufacturability (PlottedValues = Θ_{10}, Θ_{20})

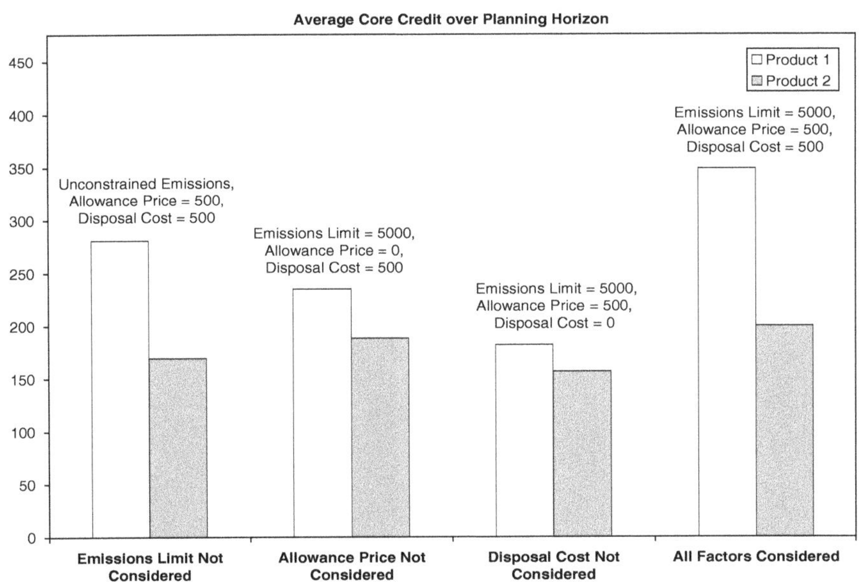

Fig. 14.6 Effect of environmental factors on core credit decision (Plotted Values = $\frac{1}{10}\sum_{t=1}^{10}\Psi_{1t}$, $\frac{1}{10}\sum_{t=1}^{10}\Psi_{2t}$)

References

Bloemhof-Ruwaard JM, van Beek P, Hordijk L, Van Wassenhove LN (1995) Interactions between operational research and environmental management. Eur J Oper Res 85:229–243

EC (European Commission) (2001) Green paper on integrated product policy

EC (European Commission) (2011) Policies. Retrieved from http://ec.europa.eu/environment/policy_en.htm. Accessed 30 Jan 2011

EPA (Environmental Protection Agency), United States (2005) Acid Rain Program: 2004 Progress Report

Environmental Resources Management (2005) Study and assessment of available information for a pilot project on a teak garden chair: final report. Oxford, UK

Fourer R, Gay DM, Kernighan BW (2003) AMPL: A modeling language for mathematical programming. Brooks/Cole-Thomson Learning, California

Fourer R (2011) Nonlinear programming frequently asked questions. Retrieved from http://158.193.150.10/nonlinear-programming-faq.html. Accessed 30 Jan 2011

NEOS (Network-Enabled Optimization System) (2011) Optimization software guide. Retrieved from http://www.neos-guide.org/NEOS/index.php/Optimization_Software_Guide. Accessed 30 Jan 2011

Nokia Corporation (2005) Integrated product policy pilot project life cycle environmental issues of mobile phones. Espoo, Finland

Scheer D, Rubik F (eds) (2005) Governance of integrated product policy: in search of sustainable production and consumption. Greenleaf Publishing Ltd., Sheffield, UK

Smith VM, Keoleian GA (2004) The value of remanufactured engines: life-cycle environmental and economic perspectives. J Ind Ecol 8(1–2):193–221

Tukker A (2006) Identifying priorities for environmental product policy. J Ind Ecol 10(3):1–4

Tukker A, Jansen B (2006) Environmental impacts of iroducts: a detailed review of studies. J Ind Ecol 10(3):159–182

Chapter 15
Performance Measurement and Evaluation for Sustainable Supply Chains using Rough Set and Data Envelopment Analysis

Chunguang Bai and Joseph Sarkis

Abstract Performance measurement of sustainable supply chains is not a trivial issue. The complexities associated with measurement of supply chains is well known. Expanded sustainability measures for organizations and their supply chains only causes additional complexity. With many and varied measures helping organizations to distill information such that only the most pertinent and direct measures that provide information can make the process both more effective and efficient. In order to meet these goals this chapter provides an integrative approach that will help first distill and filter measures that are less redundant and then utilize these measures to arrive at a single performance value that managers can use for comparative and benchmarking analysis. We utilize a novel rough set theoretic approach and data envelopment analysis (DEA) as tools for distillation and integration of a variety of sustainable supply chain performance measures. An example from sustainable supply chain data helps provide insights into the methodology.

Evaluating and monitoring supply chain management programs is dependent on the development and application of performance measures. This evaluation becomes more complex when considering the sustainability of these supply chains. The use of performance measurement for sustainable supply chain management can be used to select suppliers, monitor performance of suppliers, and for general supplier development purposes. The use of performance measurement systems for sustainable supplier evaluation purposes has been recommended (Hervani et al. 2005). A critical aspect of this performance measurement system is the identification of appropriate measures that may be used for evaluation. Given that there may be dozens, and sometimes hundreds, of measures that can be used for supplier evaluation and even more so when additional dimensions of sustainability are included in the evaluation, the need to identify the most appropriate measures becomes more critical. There are

C. Bai (✉)
School of Management Science and Engineering, Dongbei University of Finance and Economics, 217 Jianshan Street, 116025 Dalian, People's Republic of China
e-mail: chunguang.bai@gmail.com

J. Sarkis
Graduate School of Management, Clark University, 950 Main Street, 01610-1477, Worcester, MA, USA
e-mail: jsarkis@clarku.edu

T. Boone et al. (eds.), *Sustainable Supply Chains*,
International Series in Operations Research & Management Science 174,
DOI 10.1007/978-1-4419-6105-1_15,

techniques that have been developed and applied to help reduce the performance measurement set to a set of "key performance indicators". These key performance indicators may or may not lose information from the full set of indicators. The use of information theoretic tools, such as rough set theory, may be valuable to determine a usable subset of key performance indicators for supplier evaluation.

Another dimension of supplier evaluation is through various benchmarking tools. One of the more familiar tools for supplier evaluation is data envelopment analysis (DEA). Even though DEA has been utilized for supplier performance evaluation (Talluri and Sarkis 2002), its application to sustainable supplier evaluation has been more limited. DEA is very much dependent on input and output performance measurement data. Two primary issues facing the application of DEA is the appropriate selection of input and output metrics, and making sure that a useful data set exists (Sarkis 2007). DEA is also quite sensitive to the data used in the evaluation. Thus, reducing performance data information through information theoretic approaches such as rough set analysis, may cause variations in the solution to relative performance of sustainable supply chain partners. This chapter will investigate this sensitivity using an illustrative sustainable supply chain data set and a neighborhood rough set technique. The level of comparability of results based on reduced performance data will be evaluated with some implications identified.

15.1 Background

As part of the background discussion, we briefly introduce the importance of sustainable supply chain performance measurement, an introduction into the rough set and neighborhood rough set technique, and also into the DEA performance approach.

15.1.1 Sustainable Supply Chain Performance Measurement

The number of performance measures that can be utilized for evaluation of a supply chain can be quite extensive. One of the more popular and extensive performance evaluation systems is the supply chain operations reference (SCOR) model. The SCOR model categorizes the processes of five supply chain stages including plan, source, make, deliver and return. We focus on the source function in our study due to this function's direct activities with suppliers. The performance measures within SCOR are categorized on Cost, Time, Quality, Flexibility, and Innovation dimensions (Shepherd and Gunter 2006). Traditionally, these performance measures are business oriented, environmental sustainability measures are lacking.

Given both business and environmental sustainability measures, the set of performance measures can be quite extensive. The number of performance measures that can be used to evaluate just one set of relationships within the supply chain can become quite extensive. For a listing of example performance measures for a supply chain and sustainability within the supply chain are shown in Table 15.1.

Table 15.1 SCOR-based sourcing function business and environmental performance measures categorized on cost, time, quality, flexibility, and innovation dimensions. (Talluri and Sarkis 2002; Shepherd and Gunter 2006; Gunasekaran and Kobu 2007)

Performance measures	
Business sustainability	Environmental sustainability
Cost	
Supplier cost-saving initiatives Labor efficiency Cost variance from expected costs	Environmental costs savings Energy efficiency of systems Environmental cost performance variance Amount of environmental penalties
Time	
Supplier lead time against industry norm Supplier's booking-in procedures Purchase order cycle time Percentage of late deliveries Information timeliness Efficiency of purchase order cycle time	Length to time to implement environmental programs Meeting environmental program implementation period Speed of acquiring environmental information Communication speed on environmental issues to supplier's suppliers
Quality	
Buyer-supplier partnership level Level of supplier's defect-free deliveries Supplier rejection rate Delivery reliability Percentage of wrong supplier delivery Mutual trust Satisfaction with knowledge transfer Satisfaction with supplier relationship Supplier assistance in solving technical problems Extent of mutual planning cooperation leading to improved quality Extent of mutual assistance leading in problem-solving efforts Distribution of decision competences between supplier and customer Quality and frequency of exchange of logistics information between supplier and customer Quality of perspective taking in supply networks Information accuracy Information availability	Environmental relationship and cooperation level Supplier rejection rate Waste generated from products and materials Percentage recycled material Mutual trust on environmental issues Mutual planning for environmental improvements Mutual assistance for environmental improvements Environmental information accuracy Environmental information availability
Flexibility	
Supplier ability to respond to quality problems Response to product changes Materials variety (number of materials available) Product and service variety Product volume variability capabilities Product development time	Amount of environmentally safe alternatives Response to environmental programs for suppliers Response to environmental product requests

Table 15.1 (continued)

Performance measures	
Business sustainability	Environmental sustainability
Innovation	
Satisfaction with knowledge transfer satisfaction	Environmental knowledge Transfer satisfaction
Technological capability levels	Environmental technology Levels
Involvement in new product design	New environmentally sound processes introduced
Introduction of new processes	New environmentally sound product development

Industry is always seeking to become more efficient. Having cumbersome and complex supply chain performance measures diminishes management efficiency of the supply chain. Thus, the identification and application of key performance indicators (KPI) becomes even more valuable to supply chain managers (Chae 2009).

A major element within sustainable supply chain management is concerned with inter-organizationally sharing responsibility for various aspects of environmental and social performance. Within the environmental arena several techniques exist to help managers map the environmental impacts along supply chains, such as the life cycle assessment, product stewardship, and design for environment (DFE) principles. Performance measurement and metrics are critical to all these dimensions of life cycle assessment.

15.1.2 Rough Set Theory and Neighborhood Rough Sets

To help reduce the number of performance measures used in a study, we rely on neighborhood rough set methodology. This methodology fits within the family of rough set theory. Rough set theory is based on information theory to help reduce the number of factors with applications primarily within the data mining realm.

Rough set theory (Pawlak 1982) method classifies objects into similarity classes containing objects that are indiscernible with respect to previous occurrences and knowledge (information). Rough set methodology was developed as a non-parametric data-mining approach that can effectively determine core relationships amongst a variety of factors. It has been utilized for diverse applications as investigating marketing data (Shyng et al. 2007), education, (Beynon et al. 2000), credit risk for financial information (Ong et al. 2005), manufacturing process control (Sadoyan et al. 2006). The supply chain literature's application of rough set is only a recent occurrence. Recently application of rough set for sustainable and green supply chain and operations management concerns has occurred (Bai and Sarkis 2010).

Rough set theory is advantageous to other approaches for data-mining that typically utilize multivariate statistics that require specific parametric assumptions. It can integrate both tangible and intangible information.

In neighborhood rough sets the determination of what should be included in a reduced rough set of performance measures is allowed to have some flexibility based on a distance relationship for the attributes, which expands the boundary dimensions of a rough set. Neighborhood rough set models can deal directly with continuous numeric data, whereas traditional rough set typically requires discretization or categorization of continuous data to be effective. Use of actual continuous numeric data lessens the likelihood of information loss due to categorizing data. The definitions and methodology presented in this paper for neighborhood rough sets are based on the developments of (Hu et al. 2008a, b).

Definition 1 Given an arbitrary $x_i \in U$ and $B \subseteq C$, the neighborhood $\delta_B(x_i)$ of x_i in attribute space B is defined as

$$\delta_B(x_i) = \{x_j | x_j \in U, \Delta^B(x_i, x_j) \leq \delta\}, \tag{15.1}$$

where Δ is a distance function. For $\forall x_1, x_2, x_3 \in U$, and satisfies the following four conditions:

$$\begin{aligned} &(1)\ \Delta(x_1, x_2) \geq 0; \\ &(2)\ \Delta(x_1, x_2) = 0 \ \textit{if and only if}\ x_1 = x_2; \\ &(3)\ \Delta(x_1, x_2) = \Delta(x_2, x_1); \\ &(4)\ \Delta(x_1, x_3) \leq \Delta(x_1, x_2) + \Delta(x_2, x_3); \end{aligned} \tag{15.2}$$

Let v_{ij} denote the value of object i on attribute j, then a Minkowsky distance function may be defined as:

$$\Delta_p(x_i, x_k) = \left(\sum_{j=1}^{N} \left| v_{ij} - v_{kj} \right|^p \right)^{1/p} \tag{15.3}$$

when $p=1$ it is defined as a Manhattan distance; $p=2$ is a Euclidean distance; and $p=\infty$ is a Chebychev distance (Wilson and Martinez 1997). The Chebychev distance equation is represented by expression (15.4):

$$\Delta_\infty(x_i, x_k) = \max_j \left| v_{ij} - v_{kj} \right| \tag{15.4}$$

Given that we can identify the most salient performance measures through Rough set and neighborhood rough set theory, we can now utilize a tool for performance evaluation of multiple factors defined as data envelopment analysis (DEA).

15.1.3 Data Envelopment Analysis

DEA is a multifactor productivity measurement tool based on multiple inputs and/or outputs for a given decision making unit (DMU). DEA-based models addressing a variety of problem environments have been proposed in the literature. For the purposes of our study, a DMU is defined as an electricity generating plant.

Tone (2002) provides an important distinction between two types of measures or approaches in DEA. These measures are termed *radial* and *non-radial* measures respectively based on differences in the relationships of input or output items to each other and whether it is based on ratio or slack based modeling approaches (also additive versus ratio based).

The original DEA basic multifactor productivity models introduced by Charnes, Cooper and Rhodes (Charnes et al. 1978) (CCR), (also defined as the constant returns to scale ratio-based radial model), and by Banker (1984) (BCC) (also defined as the variable returns to scale ratio-based radial model), effectively considered multiple input and multiple output factors in evaluating relative efficiencies of DMU. Even though there are advantages and disadvantages with various forms of DEA models, we will focus the slack based models for evaluation in this chapter (see Cooper et al. 2007 for more details). We describe one particular model in more detail.

15.1.4 A General DEA Super-Efficiency Slack Based Model

Even though many different DEA models exist we will utilize the Tone (Tone 2002) super efficiency slack based input oriented model with constant returns to scale (Super SBM-I-C) for evaluation of sustainable supply chain performance. The model is formulated as:

(SUPER SBM-I-C)

$$\delta^* = \min \delta = \frac{1}{m}\sum_{i=1}^{m}\frac{\bar{x}_i}{x_{io}}$$

subject to:

$$\bar{\mathbf{x}} \geq \sum_{j=1,\neq o}^{n} \lambda_j \mathbf{x}_j$$

$$\bar{\mathbf{y}} \leq \sum_{j=1,\neq o}^{n} \lambda_j \mathbf{y}_j$$

$$\bar{\mathbf{x}} \geq \mathbf{x}_o$$

$$\bar{\mathbf{y}} = \mathbf{y}_o$$

$$\lambda \geq 0$$

where:

$\mathbf{x}_j$ = Are the vector of input factor values for supplier j
$\mathbf{y}_j$ = full set of output factor values, for supplier j

$\mathbf{x}_o$ = the 'test' Supplier input factor values
$\mathbf{y}_0$ = the 'test' Supplier output factor values
m = number of inputs
n = number of DMUs (suppliers)
λ = envelopment (facet) weight for a specific supplier solution

15.2 An Illustrative Application

Given the definitions in the previous section on neighborhood rough set characteristics and expressions, we now introduce a multi-step procedure within the context of an illustrative application. There are 5 steps in the rough set portion of the two-staged methodology to evaluate supplier sustainability performance. To set the stage of the application, we will focus on the supply chain activities associated with the SCOR 'sourcing' function. Further details of the illustrative application will occur in the steps defined below (additional information and detail can be provided in Bai and Sarkis (2012).

Step 1: Construct the Original Neighborhood Information System (Tables 15.2 and 15.3) To be able to construct a decision table, we need to identify the various performance measures and factors utilized within the illustrative example. The five SCOR performance categories are: Cost, Time, Quality, Flexibility and Innovation. The exemplary measures and the metrics of the 'raw' data are shown in Tables 15.2 and 15.3. Table 15.2 contains the input performance factors[1] of DEA data, while Table 15.3 provides examples of the output performance factors of DEA data.

Historical extreme or ideal ranges from best to worst are summarized in following list which contains input and output performance measures and their definitions and metrics used:

a. *Input performance measures*
   - Cost:
     - Cost_IN: Environmental cost performance variance*, cost variance as percentage of expected cost (0–50%)
   - Time:
     - Time1_IN: Supplier's booking-in procedures, number of booking-in procedures (range:1–10)
     - Time2_IN: Purchase order cycle time*, purchase order cycle time in days (1–60 days)
     - Time3_IN: Percentage of late deliveries, percentage of late deliveries (0–50%)

[1] An input performance measure is defined as a measure that improves as its value decreases. An output performance measure is defined as a measure that improves as its value increases (Sarkis 2000).

Table 15.2 Evaluation of suppliers on performance measurement by decision makers. (DEA INPUT values)

	Cost_IN (%)	Time1_IN	Time2_IN (day)	Time3_IN (%)	Quality1_IN (%)	Quality2_IN (%)	Flexibility_IN (day)	Innovation_IN (month)
Supplier1	45	6	45	35	9.00	1	2	19
Supplier2	21	9	31	1	7.00	9	53	13
Supplier3	16	4	17	39	5.00	4	41	14
Supplier4	7	9	52	35	4.00	8	54	8
Supplier5	32	6	39	30	4.00	10	11	14
Supplier6	23	3	50	21	3.00	0	26	1
Supplier7	42	2	37	9	7.00	4	41	17
Supplier8	37	7	39	29	6.00	10	16	23
Supplier9	44	8	45	31	3.00	4	16	7
Supplier10	22	9	5	41	4.00	7	20	12
...	...	...	...	...	...	...	...	...
Supplier116	28	5	51	4	10.00	4	43	18
Supplier117	33	8	59	38	4.00	1	46	5
Supplier118	8	2	12	6	9.00	10	22	15
Supplier119	5	10	40	25	6.00	1	19	13
Supplier120	6	7	6	43	7.00	5	42	18

Table 15.3 Evaluation of suppliers on performance measurement by decision makers. (DEA OUTPUT)

	Cost1 _OUT	Cost2 _OUT	Cost3 _OUT (%)	Time1 _OUT (%)	Time2 _OUT	Time3 _OUT (%)	Quality1 _OUT	Quality2 _OUT	Quality3 _OUT	Quality4 _OUT	Quality5 _OUT	Flexibility1 _OUT	Flexibility2 _OUT	Flexibility3 _OUT	Innovation1 _OUT	Innovation2 _OUT	Innovation3 _OUT	Innovation4 _OUT
Supplier1	2	4	13	−7	L	40	H	M	L	L	L	L	27	60	L	VL	L	3
Supplier2	7	9	11	24	L	48	H	H	M	H	M	VH	28	257	H	H	L	0
Supplier3	3	0	41	5	VL	18	H	H	M	L	H	H	18	478	L	H	M	2
Supplier4	5	8	27	12	M	24	M	VL	VL	H	VL	H	20	189	H	VL	H	2
Supplier5	1	1	41	8	L	11	L	VH	L	VL	L	L	19	26	L	H	H	2
Supplier6	6	5	1	−20	H	9	H	VH	M	M	M	L	22	495	VH	VL	VH	2
Supplier7	1	8	22	−18	VL	42	VL	M	M	M	M	VL	41	104	M	M	L	2
Supplier8	2	8	17	−1	H	40	H	VH	M	M	VH	VH	40	204	H	L	L	3
Supplier9	5	7	6	7	M	43	L	H	H	VL	VH	L	33	135	L	VH	VL	3
Supplier10	7	7	30	14	M	38	H	VH	M	H	H	M	48	307	M	L	L	4
. . .	. . .	. . .	. . .	. . .	. . .	. . .	. . .	. . .	. . .	. . .	. . .	. . .	. . .	. . .	. . .	. . .	. . .	. . .
Supplier116	7	5	45	17	L	30	M	VL	H	L	M	H	20	16	VH	L	L	0
Supplier117	4	5	35	−16	H	43	M	H	VL	L	L	VL	28	468	L	H	L	1
Supplier118	2	9	33	−16	H	16	M	H	L	H	M	L	43	358	H	VH	M	4
Supplier119	9	2	22	−23	VL	18	L	VH	L	L	M	L	49	439	M	H	H	2
Supplier120	10	5	40	−4	L	30	M	H	L	VL	VH	L	39	3	M	VL	VL	3

  - Quality:
    - Quality1_IN: Supplier rejection rate* (Qual1), percentage of materials rejected from suppliers based on environmental norm (0.01–10%)
    - Quality2_IN: Percentage of wrong supplier delivery, (0–10%)
  - Flexibility:
    - Flexibility_IN: Response to environmental product requests, responding time in days (0–60 days)
  - Innovation:
    - Innovation_IN: Product development time, product time in months (0–24 months)

b. *Output performance measures*
  - Cost:
    - Cost1_OUT: Environmental costs savings initiatives* (range: 0–10 initiatives)
    - Cost2_OUT: Number of Innovations (range: 0–10 initiatives)
    - Cost3_OUT: Energy Efficiency of systems, percentage of Production compare Energy (range: 0–50%)
  - Time:
    - Time1_OUT: Supplier lead time against industry norm* (Time1), percentage over or under industry average lead time (+25% to −25%)
    - Time2_OUT: Speed of acquiring environmental information, speed level based on following categories Very High (VH), High (H), Moderate (M), Low (L), and Very Low (L)
    - Time3_OUT: Efficiency of purchase order cycle time, percentage of purchase order cycle time (0–50%)
  - Quality:
    - Quality1_OUT: Level of supplier's defect-free deliveries, defect-free deliveries level based on following categories Very High (VH) to Very Low (L)
    - Quality2_OUT: Delivery reliability, reliability level based on following categories Very High (VH), High (H), Moderate (M), Low (L), and Very Low (L)
    - Quality3_OUT: Extent of mutual planning cooperation leading to improved quality*(Qual2), cooperation level based on following categories Very High (VH), to Very Low (L)
    - Quality4_OUT: Mutual assistance for environmental improvements, assistance level based on following categories Very High (VH), to Very Low (L)
    - Quality5_OUT: Environmental information accuracy, level based on following categories Very High (VH), High (H), Moderate (M), Low (L), and Very Low (L)
  - Flexibility:
    - Flexibility1_OUT: Supplier ability to respond to quality problems* (Flex1), ordinal qualitative evaluation of response from VH to VL

- Flexibility2_OUT: Product and service variety, number of product and service(1–50)
- Flexibility3_OUT: Materials variety (number of materials available)* (Flex2), number of materials (1–500)

- Innovation:
  - Innovation1_OUT: Environmental knowledge transfer satisfaction* (Innv1), level of satisfaction with knowledge transfer VH to VL
  - Innovation2_OUT: Technological capability levels, level of technological capability VH to VL
  - Innovation3_OUT: New environmentally sound product development, level of development VH to VL
  - Innovation4_OUT: Introduction of new processes* (Innv2), number of new processes introduced on an annual basis (0–4).

The objects in this neighborhood decision system are represented by 120 suppliers. For the rough set model the 'conditional attributes' include 8 input performance measures and 18 output performance measures and 120 suppliers for the information table. When these two tables are integrated we have 26 total conditional attributes (made up of the 8 input and 18 output performance measures) and 120 objects in the rough set 'information system'. In this illustrative example we will keep the input and output information tables together, narrowing down the number of performance measures overall. We also run one model with the input and output sets of attributes reduced separately through the rough set approach later in this chapter. The definitions of the measures and the metrics used to measure these various conditional attributes (performance measures) are shown above.

Step 2: Develop an Adjusted Neighborhood Information System. (Table 15.4) For consistency in the evaluations we introduce a normalization procedure such that the metric values for each of the conditional attributes and all the later calculations, e.g., distance and threshold measures, are using similar scales. This normalization will adjust all the conditional attribute (performance measurement) values (v_{ij}) to be $0 \leq v_{ij} \leq 1$.

We will complete a simple linear normalization based on expression (15.5).

$$\tilde{v}_{ij} = \frac{\left| v_{ij} - v_j^{\min} \right|}{\left| v_j^{\max} - v_j^{\min} \right|} \tag{15.5}$$

Where $\tilde{v}_{ij}$ is the normalized value for supplier i, performance measure j; $v_j^{\max}$ is the maximum historical value of performance measure j; $v_j^{\min}$ is the minimum historical value of performance measure j. The maximum and minimum for each performance measure is based on historical extreme ideal and nadir points and ranges. For example, for the Cost_IN metric the minimum historical value is Cost Variance as 0% of expected cost, the maximum is 50%, thus the extreme historical range is 0–50% (appearing as a denominator). Supplier 1 has a performance metric of 45%, thus the result is a normalized score of 0.9. The normalized results are shown in Table 15.4.

Table 15.4 Adjusted scores of suppliers on performance measurement

	Cost_IN	Time1_IN	Time2_IN	Time3_IN	Quality1 1_IN	Quality2 1_IN	Flexibility_IN	Innovation_IN	Cost1_OUT	Cost2_OUT	Cost3_OUT	Time1_OUT	Time2_OUT	Time3_OUT	Quality1_OUT	Quality2_OUT	Quality3_OUT	Quality4_OUT	Quality5_OUT	Flexibility1_OUT	Flexibility2_OUT	Flexibility3_OUT	Innovation1_OUT	Innovation2_OUT	Innovation3_OUT	Innovation4_OUT
Supplier1	0.9	0.56	0.75	0.7	0.9	0.1	0.03	0.79	0.2	0.4	0.26	0.36	0.4	0.8	0.8	0.6	0.4	0.4	0.4	0.4	0.53	0.12	0.4	0.2	0.4	0.75
Supplier2	0.42	0.89	0.51	0.02	0.7	0.9	0.88	0.54	0.7	0.9	0.22	0.98	0.4	0.96	0.8	0.8	0.6	0.8	0.6	1	0.55	0.51	0.8	0.8	0.4	0
Supplier3	0.32	0.33	0.27	0.78	0.5	0.4	0.68	0.58	0.3	0	0.82	0.6	0.2	0.36	0.8	0.8	0.6	0.4	0.8	0.8	0.35	0.96	0.4	0.8	0.6	0.5
Supplier4	0.14	0.89	0.86	0.7	0.4	0.8	0.9	0.33	0.5	0.8	0.54	0.74	0.6	0.48	0.6	0.2	0.2	0.8	0.2	0.8	0.39	0.38	0.8	0.2	0.8	0.5
Supplier5	0.64	0.56	0.64	0.6	0.4	1	0.18	0.58	0.1	0.1	0.82	0.66	0.4	0.22	0.4	1	0.4	0.2	0.4	0.4	0.37	0.05	0.4	0.8	0.8	0.5
Supplier6	0.46	0.22	0.83	0.42	0.3	0	0.43	0.04	0.6	0.5	0.02	0.1	0.8	0.18	0.8	1	0.6	0.6	0.6	0.4	0.43	0.99	1	0.2	1	0.5
Supplier7	0.84	0.11	0.61	0.18	0.7	0.4	0.68	0.71	0.1	0.8	0.44	0.14	0.2	0.84	0.2	0.6	0.6	0.6	0.6	0.2	0.82	0.21	0.6	0.6	0.4	0.5
Supplier8	0.74	0.67	0.64	0.58	0.6	1	0.27	0.96	0.2	0.8	0.34	0.48	0.8	0.8	0.8	1	0.6	0.6	1	1	0.8	0.41	0.8	0.4	0.4	0.75
Supplier9	0.88	0.78	0.75	0.62	0.3	0.4	0.27	0.29	0.5	0.7	0.12	0.64	0.6	0.86	0.4	0.8	0.8	0.2	1	0.4	0.65	0.27	0.4	1	0.2	0.75
Supplier10	0.44	0.89	0.07	0.82	0.4	0.7	0.33	0.5	0.7	0.7	0.6	0.78	0.6	0.76	0.8	1	0.6	0.8	0.8	0.6	0.96	0.61	0.6	0.4	0.4	1
Supplier116	0.56	0.44	0.85	0.08	1	0.4	0.72	0.75	0.7	0.5	0.9	0.84	0.4	0.6	0.6	0.2	0.8	0.4	0.6	0.8	0.39	0.03	1	0.4	0.4	0
Supplier117	0.66	0.78	0.98	0.76	0.4	0.1	0.77	0.21	0.4	0.5	0.7	0.18	0.8	0.86	0.6	0.8	0.2	0.4	0.4	0.2	0.55	0.94	0.4	0.8	0.4	0.25
Supplier118	0.16	0.11	0.19	0.12	0.9	1	0.37	0.63	0.2	0.9	0.66	0.18	0.8	0.32	0.6	0.8	0.4	0.8	0.6	0.4	0.86	0.72	0.8	1	0.6	1
Supplier119	0.1	1	0.66	0.5	0.6	0.1	0.32	0.54	0.9	0.2	0.44	0.04	0.2	0.36	0.4	1	0.4	0.4	0.6	0.4	0.98	0.88	0.6	0.8	0.8	0.5
Supplier120	0.12	0.67	0.08	0.86	0.7	0.5	0.7	0.75	1	0.5	0.8	0.42	0.4	0.6	0.6	0.8	0.4	0.2	1	0.4	0.78	0	0.6	0.2	0.2	0.75

Step 3: Compute Neighborhood Relational Matrices for the Performance Measures (Conditional Attributes) Neighborhood relations are a form of similarity relationship. Neighborhood relations are used to group sets of suppliers (objects) based on similarity or indistinguishability in terms of distances using the neighborhood distance parameter. Our first sub-step in this process is to calculate the distances between each of the suppliers on each of the conditional attributes. This will result in a 26 n $\times$ n distance matrices for the input measures. Since initially there is only 1 attribute, we only consider the simple distance between two suppliers based on the attribute. If more than one attribute exists we may wish to use the Chebychev distance calculation as shown in expression (15.4). For example, for the Cost_IN conditional attribute the distance between Supplier1 and Supplier2 is simply $|0.9-0.42| = 0.48$. For the sake of brevity we do not show the matrix with the distance scores.

The next sub-step determines the neighborhood relation for each performance measure we need to arrive at a relational matrix. The neighborhood relational matrix is defined as:

$$M^B(N) = (r_{ij})_{n \times n}, \quad \text{where} \quad r_{ij} = \begin{cases} 1, & \Delta(x_i, x_j) \leq \delta, \\ 0, & otherwise. \end{cases} \tag{15.6}$$

For our illustrative example we begin with a neighborhood granule where $\delta = 0.2$.

All the diagonal elements of this matrix are equal to 1. That is $r_{ii} = 1$ because $\Delta^B(x_i, x_i) = 0 \leq \delta$.

In the initial example matrix, the set B is a reduct set that is the union of the performance metric Cost_IN and the previous reduct set *Atr.* In the initialization step the core performance measurement in our reduct set is set to $Atr = \varnothing$. Thus, set B is just the individual performance metrics that we are calculating.

For the performance metric Cost_IN we know that $r_{12} = r_{21} = 0$ because $\Delta^{Cost\_IN}(x_1, x_2) = 0.420 \geq 0.2$. $M^{Cost\_IN}(120) = (r_{ij})_{120 \times 120}$ is shown in Table 15.5; an additional 25 relational matrices (not shown here) can be determined for each of the performance measures separately.

In later steps $B = \{a_j \bigcup Atr | a_j \notin Atr\}$. Where *Atr* is the core performance measurement set (performance measures) and a_j is from the set of unselected attributes (performance measures). Initially, $Atr = \varnothing$ We return to Step 3 after Step 6 until no additional significant attributes are found (i.e.$Sig_I(a_j, B) \leq \varepsilon$). Each cycle of these steps will result in identifying only one additional significant attribute. The final *Atr* core performance measurement set is the reduced outcome.

Step 4: Determine the Information Significance of a Performance Measure The first sub-step in this process is to compute the neighborhood information granule ($\delta_B(x_i)$; see Definition 1) for each of the suppliers for a given performance measure or set of performance measures B.

To calculate the information significance of a performance measure j (a_j) with respect to a set B, we use expression (15.7).

$$Sig(a_j, B) = 1 - \frac{1}{|U|^2} \sum_{i=1}^{|U|} |\delta_B(x_i)| \tag{15.7}$$

Table 15.5 The relational matrix for the COST_IN performance measurement between suppliers' normalized scores

No.	1	2	3	4	5	6	7	8	9	10	11	12	13	14	15	106	107	108	109	110	111	112	113	114	115	116	117	118	119	120
1	1	0	0	0	0	0	1	1	1	0	0	0	1	1	0	1	1	0	1	0	0	0	0	0	0	0	0	0	0	0
2	0	1	1	0	0	1	0	0	0	1	1	0	0	0	1	0	0	0	0	1	1	0	1	0	1	1	0	0	0	0
3	0	1	1	1	0	1	0	0	0	1	1	0	0	0	1	0	0	0	0	0	1	0	1	1	1	0	0	1	0	0
4	0	0	1	1	0	0	0	0	0	0	0	1	0	0	0	0	0	0	0	0	0	1	0	1	0	0	0	1	1	1
5	0	0	0	0	1	1	0	1	0	0	0	0	1	1	0	0	0	1	1	1	0	0	0	0	1	1	1	0	0	0
6	0	1	1	0	1	1	0	0	0	1	1	0	0	0	1	0	0	1	0	1	1	0	1	0	1	1	0	0	0	0
7	1	0	0	0	0	0	1	1	1	0	0	0	1	1	0	1	1	0	1	0	0	0	0	0	0	0	1	0	0	0
8	1	0	0	0	1	0	1	1	1	0	0	0	1	1	0	0	1	1	1	1	0	0	0	0	0	1	1	0	0	0
9	1	0	0	0	0	0	1	1	1	0	0	0	1	1	0	1	1	0	1	0	0	0	0	0	0	0	0	0	0	0
10	0	1	1	0	0	1	0	0	0	1	1	0	0	0	1	0	0	1	0	1	1	0	1	0	1	1	0	0	0	0
11	0	1	1	0	0	1	0	0	0	1	1	0	0	0	1	0	0	0	0	1	1	0	1	0	1	1	0	0	0	0
12	0	0	0	1	0	0	0	0	0	0	0	1	0	0	0	0	0	0	0	0	0	1	0	1	0	0	0	1	1	1
13	1	0	0	0	1	0	1	1	1	0	0	0	1	1	0	0	1	1	1	1	0	0	0	0	0	1	1	0	0	0
14	1	0	0	0	1	0	1	1	1	0	0	0	1	1	0	0	1	1	1	0	0	0	0	0	0	0	1	0	0	0
15	0	1	1	0	0	1	0	0	0	1	1	0	0	0	1	0	0	0	0	1	1	0	1	0	1	1	0	0	0	0
...	...	...	...	...	...	...	...	...	...	...	...	...	...	...	...	...	...	...	...	...	...	...	...	...	...	...	...	...	...	...
106	1	0	0	0	0	0	1	0	1	0	0	0	0	0	0	1	1	0	1	0	0	0	0	0	0	0	0	0	0	0
107	1	0	0	0	0	0	1	1	1	0	0	0	1	1	0	1	1	0	1	0	0	0	0	0	0	0	0	0	0	0
108	0	0	0	0	1	1	0	1	0	1	0	0	1	1	0	0	0	1	1	1	0	0	0	0	1	1	1	0	0	0
109	1	0	0	0	1	0	1	1	1	0	0	0	1	1	0	1	1	1	1	0	0	0	0	0	0	0	1	0	0	0
110	0	1	0	0	1	1	0	1	0	1	1	0	1	0	1	0	0	1	0	1	0	0	1	0	1	1	1	0	0	0
111	0	1	1	0	0	1	0	0	0	1	1	0	0	0	1	0	0	0	0	0	1	0	1	1	1	0	0	1	0	0
112	0	0	0	1	0	0	0	0	0	0	0	1	0	0	0	0	0	0	0	0	0	1	0	1	0	0	0	1	1	1
113	0	1	1	0	0	1	0	0	0	1	1	0	0	0	1	0	0	0	0	1	1	0	1	0	1	1	0	0	0	0
114	0	0	1	1	0	0	0	0	0	0	0	1	0	0	0	0	0	0	0	0	1	1	0	1	0	0	0	1	1	1
115	0	1	1	0	1	1	0	0	0	1	1	0	0	0	1	0	0	1	0	1	1	0	1	0	1	1	1	0	0	0
116	0	1	0	0	1	1	0	1	0	1	1	0	1	0	1	0	0	1	0	1	0	0	1	0	1	1	1	0	0	0
117	0	0	0	0	1	0	1	1	0	0	0	0	1	1	0	0	0	1	1	1	0	0	0	0	1	1	1	0	0	0
118	0	0	1	1	0	0	0	0	0	0	0	1	0	0	0	0	0	0	0	0	1	1	0	1	0	0	0	1	1	1
119	0	0	0	1	0	0	0	0	0	0	0	1	0	0	0	0	0	0	0	0	0	1	0	1	0	0	0	1	1	1
120	0	0	0	1	0	0	0	0	0	0	0	1	0	0	0	0	0	0	0	0	0	1	0	1	0	0	0	1	1	1

Table 15.6 Overall results of neighborhood rough set methodology using various neighborhood distance values (δ)

Neighbor	Select_feature_neighbor
0	22, 3
0.05	1, 7, 21
0.1	2, 7, 6, 8
0.15	13, 19, 24, 17, 3
0.2	*26, 2, 8, 21, 1, 19*
0.25	6, 1, 9, 10, 7, 5, 11, 4
0.3	2, 22, 1, 11, 7, 21, 19, 15, 4
0.35	6, 25, 1, 19, 9, 5, 7, 22, 16, 15
0.4	26, 21, 11, 1, 2, 7, 22, 12, 14, 3, 6, 19, 4
0.45	26, 6, 7, 9, 11, 1, 3, 22, 21, 14, 12, 19, 8, 4, 15, 5
0.5	7, 21, 1, 22, 11, 14, 3, 19, 12, 6, 2, 9, 4, 15, 8, 5, 25, 24, 18, 20

In expression (15.7) $Sig(a_j, B)$ is the information significance[2] over the a_j conditional attribute, $|U|$ is the cardinality of the universe of suppliers (120 in our example). $|\delta_B(x_i)|$ is the number of suppliers with similar attributes levels across B conditional attribute for a supplier i. It is also defined as the number of members within the B conditional attribute elementary set for supplier i.

As an example, supplier 1 is in the same neighborhood value in the a_1 conditional attribute as suppliers $\{7, 8, 9, \ldots, 107\}$, thus $|\delta_B(x_1)| = 33$. The value of $|\delta_B(x_i)|$ is determined by summing the rows of $M^{Cost\_IN}(120)$.

Step 5: Select and Update Core Performance Measurement Set and Reduct This step of the methodology requires that we select the performance measure a_k that satisfies (15.8):

$$Sig(a_k, B) = \max_j(Sig(a_j, B)) \tag{15.8}$$

The performance measure with the largest significance a_k = Innovation4_OUT.

To update the core performance measure set *Atr*, the following rule is applied:

If $Sig(a_k, B) - Sig(Atr) > \varepsilon$, where ε is a positive infinitesimal real number used to control the convergence, then $Atr \bigcup a_k -> Atr$. We then return to step 3 with a new core performance measure set *Atr*. Otherwise if $Sig(a_k, B) - Sig(Atr) \leq \varepsilon$ we stop and the final reduct set and core performance measure set is *Atr*.

For our illustrative example we define $\varepsilon = 0.001$. We find that for a_k = Innovation4_OUT, Sig_1(Innovation4_OUT, Innovation4_OUT) > 0.001, so initially

$$Atr = Atr \bigcup \text{Innovation4\_OUT} = \{\text{Innovation4\_OUT}\}.$$

We then return to Step 3

After a number of iterations the final set *Atr* for $\delta = 0.2$ is: {Innovation4_OUT, Time1_IN, Innovation_IN, Flexibility2_OUT, Cost_IN, Quality5_OUT} or numerically, {26, 2, 8, 21, 1, 19}. This result is italicized and bolded in Table 15.6. We

[2] This term has also been defined as information entropy of a system (Liang and Shi 2004).

show other values of δ in this Table to provide some idea of the sensitivity of the solution.

This initial illustrative analysis shows that only 6 performance measures are needed to provide a threshold amount of information from the whole set of 26 performance measures. We now complete an evaluation using the Super SBM-I-C DEA model to determine what happens to the resulting relative performance evaluation among the various suppliers using (15.1) the whole set of performance metrics and (15.2) the 6 rough set neighborhood set we identified in this illustrative example.

15.2.1 Evaluating the Supplier Sustainability Performance Using DEA

To complete this step of the process we shall first put in all the original performance measures in the DEA model. The results of the full set of measures are shown in Table 15.7. We see that 41 is the highest ranked supplier in terms of overall sustainability performance with a DEA score of 7.510. This result may be useful to managers who wish to benchmark the processes that allow this supplier to perform well on both business and environmental dimensions. DEA also provides other information such as 'facet' sets that determine which suppliers play constraining roles in the efficiency (DEA) scores. We will not detail these in our paper, but this information may be useful for more effective sustainability benchmarking. Supplier 69 had the lowest ranking score.

The goal of this paper is to focus on the application of rough set to reduce the number of performance measures that a manager would need to collect and apply within a DEA model. Going through the five step neighborhood rough set process we determined that the most salient (provide the most complete information for the full set) were six performance measures. This reduced set of performance measures includes three input and three output measures. We now execute the DEA model and determine their scores (we do not report these scores in this chapter). The resulting rankings of the suppliers with the reduced set are shown in Table 15.7 under column headings of "Red. Rank". One interesting result is that even for the smaller set of performance measures, Supplier 41 is still ranks as the top overall sustainability performer. Supplier 69 still ranks very low using the reduced set of performance measures.

Overall, the groupings of the suppliers into high, medium and low performance are relatively consistent. The simple correlation analysis shows that it is not a perfectly matched ordering with only a 0.61 score. That means there are some significant differences.

15.3 Discussion and Conclusion

The advantages of this two-staged methodology are pretty clear, reduced data sets require less extensive acquisition of data by managers and lessened computational requirements, especially if an organization is looking at potentially thousands of

Table 15.7 DEA scores and ranks for all suppliers, including all performance measures and a reduced set of performance measures

Supplier ID	DEA score	Full rank	Red. rank	Supplier ID	DEA score	Full rank	Red. rank	Supplier ID	DEA score	Full rank	Red. rank	Supplier ID	DEA score	Full rank	Red. rank
Supplier1	1.919	29	104	Supplier31	2.241	13	8	Supplier61	2.009	23	12	Supplier91	1.096	106	97
Supplier2	1.626	49	99	Supplier32	5.491	3	51	Supplier62	1.116	105	108	Supplier92	1.695	41	73
Supplier3	1.157	102	90	Supplier33	1.225	91	111	Supplier63	1.463	64	48	Supplier93	1.627	48	62
Supplier4	1.184	98	66	Supplier34	1.577	54	86	Supplier64	1.464	63	82	Supplier94	1.663	45	84
Supplier5	0.843	119	113	Supplier35	1.432	66	77	Supplier65	1.794	35	46	Supplier95	1.551	56	63
Supplier6	4.432	4	27	Supplier36	1.626	50	18	Supplier66	1.190	97	117	Supplier96	1.023	113	118
Supplier7	1.095	107	68	Supplier37	1.634	47	34	Supplier67	1.190	96	110	Supplier97	1.466	62	7
Supplier8	0.869	117	96	Supplier38	1.519	60	69	Supplier68	1.648	46	39	Supplier98	2.648	7	30
Supplier9	1.252	90	64	Supplier39	1.762	37	47	Supplier69	0.729	120	112	Supplier99	1.125	104	89
Supplier10	1.360	74	50	Supplier40	1.428	67	92	Supplier70	1.368	72	16	Supplier100	1.139	103	107
Supplier11	1.598	51	76	Supplier41	7.510	1	1	Supplier71	1.298	85	106	Supplier101	0.911	115	100
Supplier12	1.357	75	9	Supplier42	2.120	17	41	Supplier72	1.956	26	54	Supplier102	1.351	77	28
Supplier13	1.384	70	75	Supplier43	1.182	99	79	Supplier73	1.892	32	22	Supplier103	2.225	14	55
Supplier14	1.585	53	91	Supplier44	1.211	93	98	Supplier74	1.369	71	42	Supplier104	1.170	100	119
Supplier15	1.027	112	70	Supplier45	2.065	21	38	Supplier75	1.495	61	14	Supplier105	1.288	86	26
Supplier16	1.774	36	45	Supplier46	1.586	52	25	Supplier76	1.282	87	94	Supplier106	1.917	30	44
Supplier17	0.864	118	103	Supplier47	2.082	20	72	Supplier77	1.327	82	87	Supplier107	1.331	80	61
Supplier18	2.451	11	116	Supplier48	1.091	108	67	Supplier78	1.068	110	120	Supplier108	1.303	84	83
Supplier19	3.217	5	101	Supplier49	1.921	28	4	Supplier79	2.167	15	17	Supplier109	1.325	83	65
Supplier20	1.993	24	13	Supplier50	1.190	95	78	Supplier80	0.895	116	102	Supplier110	1.202	94	32
Supplier21	2.544	9	60	Supplier51	2.156	16	6	Supplier81	1.757	38	5	Supplier111	1.331	81	93
Supplier22	1.437	65	49	Supplier52	1.907	31	37	Supplier82	1.028	111	105	Supplier112	1.694	42	24
Supplier23	1.548	57	52	Supplier53	1.575	55	10	Supplier83	2.113	18	3	Supplier113	2.347	12	20
Supplier24	1.356	76	59	Supplier54	1.389	69	43	Supplier84	2.568	8	21	Supplier114	1.809	33	57
Supplier25	6.856	2	53	Supplier55	1.693	43	31	Supplier85	1.361	73	56	Supplier115	2.869	6	71
Supplier26	1.221	92	95	Supplier56	2.064	22	11	Supplier86	1.261	89	85	Supplier116	1.070	109	115
Supplier27	1.332	79	58	Supplier57	1.395	68	35	Supplier87	2.480	10	29	Supplier117	1.272	88	114
Supplier28	2.088	19	2	Supplier58	1.963	25	74	Supplier88	1.807	34	80	Supplier118	1.922	27	15
Supplier29	1.528	59	88	Supplier59	1.165	101	40	Supplier89	1.696	40	19	Supplier119	1.663	44	33
Supplier30	1.754	39	109	Supplier60	1.005	114	81	Supplier90	1.544	58	23	Supplier120	1.337	78	36

suppliers. The use of DEA for supplier evaluation incorporating environmental and business factors has not received much attention. DEA is a valuable tool that allows the joint consideration of a number of performance measures as well as a tool for comparative analysis for suppliers. The use of rough set as an information-theoretic filtering tool for DEA has not seen much investigation either. The synergistic contributions of these two tools are valuable to managers and researchers. This is a first step in the process of evaluating how these tools can be used together and also for the purposes of evaluating sustainable supply chains.

Significant future research opportunities exist for this work. For example, we reduced data sets for a complete input/output and business/environmental set of measurements. The process may end up in a situation where one set of measurements may not remain in the analysis (e.g., all the environmental measures may be eliminated or all the output measures may have been eliminated). Thus, a separate rough set application for environmental inputs, environmental outputs, business inputs, business outputs, may be necessary to guarantee that one of each of these inputs and outputs (business/environment) remains in the final analysis. Thus, additional investigation needs to be completed to determine the sensitivity of these solutions. Another factor that we did not consider is the possibility of using different neighborhood granules (δ) to determine the sensitivity of the solution. As we see in the various evaluation of the neighborhood granules, the larger the granule value, the more performance measures that are left in after the series of rough set steps. This result means that the level of correlation between the DEA with a reduced set of measures and the complete set of measures will be higher. This proposition will also need to be investigated by future studies.

There are ample opportunities to apply this tool. Actual applications in real organizations may provide valuable insight into the feasibility of these tools and for the management of sustainable supply chains. There is significant promise in this work when seeking to implement green supply chain performance measurement systems.

Acknowledgment This work is supported by the National Natural Science Foundation of China Project (71102090).

References

Bai C, Sarkis J (2010) Integrating sustainability into supplier selection with grey system and rough set methodologies. Int J Prod Econ 124(1):252–264

Bai C, Sarkis J et al (2012) Evaluating ecological sustainable performance measures for supply chain management. Supply Chain Manage: An Int J 17(1):78–92

Banker RD (1984) Estimating most productive scale size using data envelopment analysis. Eur J Oper Res 17(1):35–44

Beynon M, Curry B et al (2000) Classification and rule induction using rough set theory. Expert Syst 17(3):136–148

Chae B (2009) Developing key performance indicators for supply chain: an industry perspective. Supply Chain Manage: An Int J 14(6):422–428

Charnes A, Cooper WW et al (1978) Measurement in the efficiency of decision making units. Eur J Oper Res 2:429–444

Cooper WW, Seiford LM, Tone K (2007) Data envelopment analysis: a comprehensive text with models, applications, references and DEA-Solver software, 2nd Edition. Springer, Heidelberg

Gunasekaran A, Kobu B (2007) Performance measures and metrics in logistics and supply chain management: a review of recent literature (1995–2004) for research and applications. Int J Prod Res 45:2819–2840

Hervani AA, Helms MM et al (2005) Performance measurement for green supply chain management. Benchmarking: An Int J 12(4):330–353

Hu Q, Yu D et al (2008a) Neighborhood rough set based heterogeneous feature subset selection. Inf Sci 178(18):3577–3594

Hu Q, Yu D et al (2008b) Neighborhood classifiers. Expert Syst Appl 34(2):866–876

Liang JY, Shi ZZ (2004) The information entropy, rough entropy and knowledge granulation in rough set theory. Int J Uncertain Fuzz 12:37–46

Ong CS, Huang JJ et al (2005) Building credit scoring models using genetic programming. Expert Syst Appl 29(1):41–47

Pawlak Z (1982) Rough sets. Int J Comput Inf Sci 11:341–356

Sadoyan H, Zakarian A et al (2006) Data mining algorithm for manufacturing process control. The Int J Adv Manuf Tech 28(3):342–350

Sarkis J (2000) A comparative analysis of DEA as a discrete alternative multiple criteria decision tool. Eur J Oper Res 123(3):543–557

Sarkis J (2007) Preparing your sata for DEA. In: Zhu ve J, Cook WD (eds) Modelling data irregulaties structural complexities in data development analysis. Springer, New York (Chapter 17)

Shepherd C, H Gunter (2006) Measuring supply chain performance: current research and future directions. Int J Prod Perform Manage 55(3/4):242–258

Shyng JY, Wang FK et al (2007) Rough Set Theory in analyzing the attributes of combination values for the insurance market. Expert Syst Appl 32(1):56–64

Talluri S, Sarkis J,(2002) A model for performance monitoring of suppliers. Int J Prod Res 40(16):4257–4269

Tone K (2002) A slacks-based measure of super-efficiency in data envelopment analysis. Eur J Oper Res 143(1):32–41

Wilson DR, Martinez TR (1997) Improved heterogeneous distance functions. J Artif Intell Res 6:1–34

Index

T. Boone et al. (eds.), *Sustainable Supply Chains*,
International Series in Operations Research & Management Science 174,
DOI 10.1007/978-1-4419-6105-1, © Springer Science+Business Media New York 2012